World Disasters Report
1996

GW00697176

International Federation
of Red Cross and Red Crescent Societies

Oxford University Press
1996

Oxford University Press, Walton Street, Oxford OX2 6DP

Oxford New York
Athens Auckland Bangkok Bombay Calcutta Cape Town Dar es Salaam Delhi Florence Hong Kong Istanbul Karachi Kuala Lumpur Madras Madrid Melbourne Mexico City Nairobi Paris Singapore Taipei Tokyo Toronto
and associated companies in Berlin Ibadan.

Oxford is a trade mark of Oxford University Press

Published in the United States by Oxford University Press Inc., New York

British Library Cataloguing-in-Publication Data: *Data available*

Library of Congress Cataloging-in-Publication Data: *Data available*

ISBN 0-19-829080-2
ISBN 0-19-829079-9 (Pbk)

Printed in Great Britain by The Bath Press Ltd, Bath, Avon.

Acknowledgements

The *World Disasters Report 1996* was edited by Nick Cater and Peter Walker.

Principal contributors: Chapter 1, Dennis Gallagher, Refugee Policy Group; Chapter 2, Ellen Messer, Brown University World Hunger Program; Chapter 3, John Mason, SCN/ACC; Jane Wallace, SCN/ACC, Judit Katona-Apte, WFP; David Alnwick, UNICEF; Box 3.4, Nick Cater; Chapter 4, Peter Walker; Siri Malchior, Danish Red Cross; Box 4.1, Oxfam; Box 4.2, Norwegian Church Aid; Chapter 5, Jo Macrae, ODI; John Borton, ODI; Chapter 6, Peter Walker; Hiroshi Higashiura, Japanese Red Cross Society; Chapter 7, Nick Cater; Box 7.3, Omar Valdimarsson; Chapter 8, Jane Morgan, American Red Cross; Chapter 9, John Sparrow; Chapter 10, Piero Calvi-Parisetti; Box 10.1, Charles Eldred-Evans; Box 10.2, Nick Cater; Box 10.3, Ian Christoplos; Chapter 11, Centre for Research on the Epidemiology of Disasters; Department of Peace and Conflict Research, Uppsala University; US Committee for Refugees. Thanks to all those who assisted contributors during travel and research.

The *World Disasters Report 1996* is financed by the International Federation of Red Cross and Red Crescent Societies with special contributions from the National Societies of Denmark, Finland, Japan, The Netherlands, and Sweden.

Contact details

International Federation of Red Cross and Red Crescent Societies
17, chemin des Crêts,
PO Box 372,
1211 Geneva 19, Switzerland
Tel: (41)(22) 730 4222
Fax: (41)(22) 730 0395
Email: secretariat@ifrc.org

Editing
Nick Cater
Words & Pictures
Tudor St Anthony, Mulchelney, Somerset TA10 0DL, England
Tel: (44)(1458) 251 727
Fax: (44)(1458) 251 749
Email: cater@ifrc.org

Photography
Magnum Photos Ltd
23-25 Old Street
London EC1V 9HL
England
Tel: (44)(171) 490 1771
Fax: (44)(171) 608 0020
Email: magnum@magnumphotos.co.uk

Contents

Burundi, 1994.
Gilles Peress/Magnum. Page 8.

Chiapas, Mexico, 1995.
Larry Towell/Magnum. Page 20.

Serbia, 1995.
Abbas/Magnum.
Page 34.

Honduras, 1987.
Gilles Peress/Magnum. Page 46.

Chechnya, 1995.
Donovan Wylie/Magnum. Page 54.

International Federation of Red Cross
and Red Crescent Societies   *inside front cover*
Acknowledgements   2
Introduction   6

Section One, Key Issues

Chapter 1 Population movements: Protecting the displaced   8
 Box 1.1 International legal standards for refugees   10
 Box 1.2 Who are internally displaced people?   11
 Box 1.3 How can environment create "refugees"?   13
  Figure 1.1 Global figures   14
  Figure 1.2 UNHCR's funding   16
  Figure 1.3 Returnee populations   19

Chapter 2 Global food supply: Food enough for the needy?   20
 Box 2.1 Food secure or insecure?   22
 Box 2.2 Mexico — food, land and rebellion   23
 Box 2.3 Food security — the regional dimension   25
  Figure 2.1 Food wars   26
  Figure 2.2 Who gets to eat?   28
  Figure 2.3 Hungry populations   29
  Figure 2.4 Micronutrient malnutrition   31
  Figure 2.5 Food-aid trends   32

Section Two, Methodologies

Chapter 3 Nutrition and food aid: Food for lives or livelihoods   34
 Box 3.1 Common micronutrient deficiency problems   37
 Box 3.2 What is an emergency?   39
 Box 3.3 Nutritional indicators   40
 Box 3.4 Displacement, technology, food — what next?   43
  Figure 3.1 Nutritional risk   35
  Figure 3.2 Calories and mortality   36
  Figure 3.3 Assistance and survival   38
  Figure 3.4 Selling food aid   42

Chapter 4 Developmental relief: Meeting more than basic needs   46
 Box 4.1 Uganda — going beyond refugee survival   48
 Box 4.2 Bosnia — rebuilding with milk and potatoes   49
 Box 4.3 Information helps ensure rights for refugees   52
 Box 4.4 Kenya — making participation work in camps   53

Section Three, The Year in Disasters 1995

Chapter 5 Aid trends: The state of the humanitarian system   54
 Box 5.1 Media and resources — making the link   56
 Box 5.2 Time to professionalise the professionals   58

Figure 5.1 The system   59
Figure 5.2 Development assistance   60
Figure 5.3 Humanitarian spending   61
Figure 5.4 Funding emergencies   62

Chapter 6 Kobe: Earthquake perceptions and survival   64
Box 6.1 Survivors' stories - "I was totally trapped"   67
Box 6.2 Health care — four phases and an earthquake   68
Box 6.3 Foreign perceptions and international response   73
Box 6.4 National response and the relief tasks
of the Japanese Red Cross   74
Figure 6.1 Deaths, injuries and losses   66
Figure 6.2 Disaster vulnerability   69
Figure 6.3 System crash   71
Figure 6.4 Reporting deaths   72

Chapter 7 Rwanda: Contradictions in crisis   76
Box 7.1 Evaluating Rwanda — the verdict, $1.4 billion later   78
Box 7.2 The military in Rwanda's disaster — should the soldiers
get off the humanitarian front line?   80
Box 7.3 Training for the future   82

Chapter 8 Oklahoma bombing: Psychological support
in disaster response   88
Box 8.1 Gathering global expertise on trauma   91

Chapter 9 Former Yugoslavia: Bringing relief to those who
must flee   98
Box 9.1 Bosnia — will there be a brain drain?   101
Box 9.2 Former Yugoslavia — the mines menace   103
Box 9.3 Watching for welfare if peace persists   105

Chapter 10 DPR Korea floods: Partnership and politics in disaster   108
Box 10.1 DPR Korea — stories of survival   111
Box 10.2 Meeting food needs today and tomorrow   113
Box 10.3 Viet Nam — unsustainable flood relief?   115

Section Four, Disasters Database

Chapter 11 Statistical analysis: Good data for effective response   118
Tables 1 and 2 Human impact by region. Annual average
over 25 years (1970-1994)   124
Tables 3 and 4 Human impact by type. Annual average
over 25 years (1970-1994)   125
Tables 5 and 6 Number of events by global region and type
over 25 years (1970-1994)   126
Tables 7 and 8 Total number of events by global region and type
for 1995   127
Table 9 Annual average numbers of people reported killed
or affected by disasters by country over 25 years (1970-1994)   128
Tables 10 and 11 Estimated annual average damage by region
and by type over five years (1990-1994) in thousands US$ 130
Table 12 Refugees and asylum seekers by country of origin   131
Table 13 Refugees and asylum seekers by host country   132
Table 14 Significant populations of internally displaced people   135
Table 15 Number of major armed conflicts by region per year
over five years (1990-1995)   136

Kobe, Japan, 1995.
Philip Jones Griffiths/ Magnum. Page 64.

Rwanda, 1995.
Sebastiao
Salgado/
Magnum.
Page 76.

Oklahoma City, USA, 1995.
Bruce Davidson/Magnum. Page 88.

Croatia, 1995.
Sebastiao
Salgado/
Magnum.
Page 98.

DPR Korea, 1988. Hiroji
Kubota/Magnum. Page 108.

Rwanda, 1994.
Sebastiao
Salgado/Magnum.
Page 118.

Table 16 Number of battle-related deaths in major armed conflicts
    per region per year over five years (1990-1995)   137
Table 17 Number of major armed conflicts by type of incompatibility
    per region per year over five years (1990-1995)   138
Table 18 Non-food emergency and distress relief, grant
    disbursements in millions US$ over ten years (1985-1994)   139
Table 19 Breakdown of food-aid deliveries by category per year
    over nine years (1987-1995) in thousand tonnes — cereals
    in grain equivalent   139

Section Five, Red Cross and Red Crescent

Bosnia, 1995.
Sebastiao
Salgado/Magnum.
Page 140.

**Chapter 12 International Conference: Putting humanitarian
    values up front**   140
    Box 12.1 What does International Humanitarian Law say?   142

Chapter 13 Code of Conduct: Governments back global standards   144

Chapter 14 National Societies: Reaching out across the world   150

Chapter 15 Delegations: The International Federation network   164
    International Federation disaster response around the world   170
        Figure 15.1 Appeals by type of disaster in 1995
            (in millions Sfr.)   170
        Figure 15.2 Appeals by region in 1995 (in millions Sfr.)   170
        Figure 15.3 Appeals by type, number of victims and value
            in 1995 in date order   171

World Disasters Report on the Internet   172
World Disasters Report information   173
World Disasters Report 1996 order form   174
Index   175
Advertisers   following 178
Fundamental Principles of the International Red Cross
and Red Crescent Movement   inside back cover

Rwanda, 1994.
James Nachtwey/Magnum. Page 144.

Azerbaijan, 1994.
Ian Berry/Magnum. Page 150.

Azerbaijan, 1994.
Ian Berry/Magnum. Page 164.

Meeting standards for survival

With the comparatively high funding levels of the early 1990s seemingly at an end, the humanitarian organisation scene is probably in for a shake-out. Survival for disaster victims and humanitarian agencies may well come to depend on whether standards of assistance and protection that disaster victims may expect from responding agencies (a victims' rights statement, as it were) can be formulated and agreed.

Developing such standards and negotiating agreement on them among all parties – disaster victims, agencies, donor governments and host governments – will be a key agency challenge during the remainder of this decade.

The International Federation of Red Cross and Red Crescent Societies is committed to play a leading role in standard-setting and believes that agencies have to take seriously their commitment to be accountable both to those who provide funds and those it is sought to assist.

The early 1990s saw tremendous growth in humanitarian action around the world.

Some agencies' spending increased 20 to 30 per cent annually, doubling or even tripling operational budgets. The numbers of people assisted rose, with agencies thinking of programmes to assist hundreds of thousands rather than, as in the past, tens of thousands of people. Emergency humanitarian spending rose from 2 per cent of total overseas development assistance at the beginning of the decade to 6 per cent today, totalling nearly 3.5 billion US dollars in 1994. But it has become clear that the boom in humanitarian assistance unleashed by the ending of the Cold War has peaked and is receding.

We may also be seeing a downturn in the amount of food aid available. Doubling in a decade, the percentage of food aid used for emergencies rather than development projects reached 40 per cent in 1995, signalling a greater targeting of a scarce resource. Projections to the year 2000 suggest that the growth in food-aid availability will not keep pace with growth in demand.

Fewer resources do not imply any reduction in numbers of people in need. As Chapter 1 of this year's *Report* shows, the world is likely to see more, not fewer, people on the move, with most fleeing civil strife. A greater proportion of these people will be trapped within their own countries and unable to seek protection as refugees. Twenty million internally displaced people were in need of assistance in 1990; by 1994 the number had grown to 26 million.

While the 1990s decade has seen an ending of some wars, particularly in Africa, others have broken out in Europe and Asia. The reshaping of nations and allegiances seems set to continue, with yet more people in flight, economic collapse and human rights violations. There will be no shortage of work for aid agencies as the decade closes.

While we can better target resources and increase efficiency and effectiveness, what of causes? Is humanitarian assistance being used as a Band-Aid to cover inaction at fundamental levels in addressing the causes of today's disasters? At the 1995 International Conference of the Red Cross and Red Crescent, some 142 governments joined in debate with the Red Cross and Red Crescent international and national components on humanitarian issues, with the Movement urging governments to address the policy vacuum.

Humanitarian assistance alone cannot address many of today's crises. Political and economic action, too, must be taken by governments to underpin effective humanitarian assistance whose independence and neutrality is respected and guaranteed. The actions of governments are needed to ensure that the components of the International Red Cross and Red Crescent Movement have the humanitarian space they need to carry out their work. Within that space, we must strive for excellence.

With the humanitarian service *Code of Conduct* as a guide, humanitarian response agencies must exceed established ethical and performance standards. A growing workload coupled with the prospect of scarcer resources means we have to become much clearer in our thinking and action. As Section Two of this *Report* shows, it is possible to deliver humanitarian assistance in ways that involve disaster victims and build rehabilitation opportunities. It is possible, for example, to think more critically about food aid, using it to build livelihoods rather than merely fill stomachs.

Agencies are going to be judged not just on presence alone, but on the quality and sustainability of their work and how accountable they are for it. Those who cannot come up to standard may soon find themselves made irrelevant on the humanitarian assistance scene. Professional response agencies must ensure effective management of change. The bottom line is effective and efficient delivery of assistance and protection to those who need it most. This ultimately is the standard against which we in the Red Cross and Red Crescent and all other humanitarian response organisations will be judged.

George Weber
Secretary General

Humanitarian framework: More people are on the move, fleeing violence and intimidation, hunger and poverty, oppression and environmental decline. Today many of them remain trapped within their own countries as internally displaced people, some flee to become refugees, some receive international assistance, others fend for themselves. Amid the ebb and flow of conflict, every move from camp to camp undermines each family's security, health and capital. The needs are enormous, the resources are not keeping pace.

Abandoned refugee camp, Burundi, 1994. Gilles Peress/Magnum

Protecting
the displaced

In the post-Cold War period, the political interest of key donor states in countries prone to complex humanitarian disasters has not been sufficient to thwart such disasters before they break out. At least to date, however, the determination of these states to avoid involvement in the problems of what they see as secondary or tertiary nations is not sufficient to prevent large-scale response to full-blown humanitarian crises, especially when the media activates substantial public interest and concern.

The irony is that key donor states and the international community often end up spending more time and resources in dealing with these emergencies than they would if they had tried to prevent them in the first place.

The result is that humanitarian budgets rise and humanitarian action begins to be viewed as a substitute for political initiative. Will key donor governments see the light and put greater energy and resources into preventing or mitigating disasters? Will they continue to do what they are now doing, which is to respond in a major way to humanitarian disasters only after they have matured and captured media attention? Or will these states constrain their response to complex humanitarian disasters, resisting more strongly media-driven reactions?

While it is not clear what the future holds for humanitarian programmes, it is clear that over the last five years efforts to help refugees and internally displaced people have grown enormously. Between 1990 and 1995 the budget of the United Nations High Commissioner for Refugees (UNHCR) increased from 500 million US dollars ($) to $1.5 billion.

Other humanitarian institutions, such as the International Red Cross and Red Crescent Movement, the World Food Programme and many non-governmental organisations (NGOs), are spending far more on refugees, displaced people and war-affected populations than they did in the recent past.

This chapter examines the factors that have led to such growth and considers the future of refugee response.

The numbers of refugees and internally displaced people have increased dramatically: there were 22 million in 1985 and, by 1995, their number had grown to 37 million. The principal causes of this increase are internal conflicts resulting from ethnic and religious tensions, and the greater number of collapsed states.

Not only have the numbers of displaced people increased but agencies now reach many more of them. In the past, humanitarian aid was often restricted to assisting refugees in asylum countries. Today, it is increasingly common for help to be given to displaced people and other war-affected populations, as well as refugees.

Until recently most of those aided by UNHCR were long-term refugees in camps; in 1985 about 80 per cent of UNHCR's resources went on care and maintenance. Today, a much higher percentage of people aided by the humanitarian community are at either the emergency or the "solutions" phase. Donor states are more willing to fund programmes that address emergency needs or promote solutions than on-going relief programmes.

These changes prompt important questions. Is the growth in funding sustainable? Does this growth distort the priorities of the international system toward a few visible emergencies? Is the growth eroding the commitment of agencies and states to humanitarian law and principles? Are humanitarian programmes used as a substitute for political action? Is an effective balance being maintained between the interests of states – particularly donor states – and the needs of refugees and displaced people? Will recent use of the military continue and under what conditions? Should the "innovative" approaches of recent emergencies – such as safe havens within conflict areas

---

Box 1.1 International legal standards for refugees

The 1951 Convention and the 1967 Protocol relating to the Status of Refugees are the core international instruments regulating the conduct of States in matters involving the treatment of refugees. The Convention, as amended by Article 1(2) of the 1967 Protocol, defines a refugee as any person who:

"Owing to a well founded fear of being persecuted for reasons of race, religion, nationality, membership of a particular social group or political opinion, is outside the country of his nationality and is unable or, owing to such fear, is unwilling to avail himself of the protection of that country; or who, not having a nationality and being outside the country of his former habitual residence as a result of such events, is unable or, owing to such a fear, is unwilling to return to it."

While states are not obligated to grant asylum, they must respect certain minimum standards for treatment of refugees. This fundamental protection prohibiting expulsion or return ("refoulement") is highlighted in the Convention, in Article 33(1):

"No Contracting State shall expel or return ("refouler") a refugee in any manner whatsoever to the frontiers of territories where his life or freedom would be threatened on account of his race, religion, nationality, membership of a particular social group or political opinion."

Regional conventions have modified the definition and accompanying protection to fit regional circumstances.

For example, the Organisation of African Unity (OAU) Convention Governing the Specific Aspects of Refugee Problems in Africa adopts a broader definition. In addition to those who meet the Convention definition:

"The term 'refugee' shall also apply to every person who, owing to external aggression, occupation, foreign domination or events seriously disturbing public order in either part or the whole of his country of origin or nationality, is compelled to leave his place of habitual residence in order to seek refuge in another place outside his country of origin or nationality."

The OAU Convention also offers guidelines for providing asylum and unambiguously stipulates that repatriation must be a voluntary act. Like the OAU Convention, the 1984 Cartegena Declaration on Refugees contains a similarly broadened definition of "refugee" to reflect the prevailing situation in the Latin American region. Although a non-binding instrument, the Declaration has been accepted and is being applied by Latin American states to the degree that it has entered the domain of international law.

While the number of uprooted people has increased, the number of those who meet the refugee definition of the 1951 Convention and its Protocol, has decreased since the end of the Cold War. Paradoxically, UNHCR would be a small office with a tiny budget if its activities were limited to Convention refugees. But its mandate has been frequently enlarged through resolutions adopted by the UN General Assembly to the point that, today, the OAU refugee definition, which includes victims of internal conflicts who sought refuge abroad, is commonly used. ∎

---

in the former Yugoslavia – be replicated? Are we already in a vicious circle where an increasingly large proportion of international aid is spent on emergencies leaving fewer resources for development or even disaster prevention and mitigation? Such questions need serious attention and analysis.

Although humanitarian action must be impartial, politics greatly influences the causes of humanitarian emergencies, which emergencies are responded to, by whom and with what resources. It influences whether humanitarian principles, laws and mandates are stressed or downplayed. It is political, not humanitarian, action that resolves the situations that cause refugees and displaced people. Humanitarian programmes must always function within the ambit of politics. What are the key political factors influencing the actions of the international humanitarian community?

While resolution of some conflicts has allowed major repatriations, ethnic, racial and religious tensions have risen in the post-Cold War power and policy vacuum. States have collapsed and external and internal borders are being redrawn. Populations are being compelled to realign themselves, producing massive numbers of refugees and internally displaced people.

Uncertainties about the stability of global and national economies are turning societies inward, and governments and their peoples exhibit less interest in the welfare of other populations. One clear indicator is the decline in support for international aid to the developing world.

Societies show greater reluctance to share their space and resources with new arrivals, especially those of a different race or religion from the majority. This includes rejection of multi-culturalism, as evidenced by tightened controls on immigration and lowering of protection for minorities. The right of

---

Box 1.2 Who are internally displaced people?

Over the last decade the numbers of internally displaced persons (IDPs) have increased exponentially. In 1990 there were an estimated 22 million IDPs; by 1994 their number had grown to about 26 million. If the current pace of increase were to continue, by the year 2000 there would be about 40 million IDPs.

There is relatively rapid turnover in IDP populations. Many IDPs flee their homes in the wake of conflict and return a few months later when the conflict recedes. These same people may be forced to flee again only months later when conflict and violence returns to their area. This means that the numbers of incidents of displacement far exceed the numbers of displaced people. It also means that the same IDP may be counted several times because he or she has been displaced more than once.

It is estimated that between 1990 and the year 2000 there will be well over 100 million separate incidences of internal displacement. Whereas in 1990, UNHCR assisted only refugees (that is, forcibly displaced people who crossed an international border), by 1995 almost half of UNHCR's case-load of approximately 25 million people was made up of IDPs and others remaining in their countries affected by internal conflicts.

Providing accurate figures on IDPs is notoriously difficult. There is a vast grey area between seasonal and economic migrants on the one hand and those fleeing war and persecution on the other.

Take, for example, China: of its rural working population of 450 million, 120 million are without real work and the number may reach 200 million by the year 2000. According to a senior Chinese official, Wu Bangguo, "If we do not properly transfer this large amount of surplus labour, this will lead not only to very large economic losses but to major social and political problems. The transient population has already caused serious problems, with many migrants turning to crime when they find there are no jobs." Of the 120 million, an estimated 80 million have already left the land to work in towns and cities across China, on building sites, in factories and restaurants and doing other jobs that urban residents do not want to do. When do these economic migrants become IDPs?

IDPs also tend to be hidden, blending in with the local population. Many may congregate in camps but a much larger number move to urban areas, to the bush or forests and/or move in with extended family or clan members. ∎

---

people to seek asylum elsewhere is downplayed as greater emphasis is placed on the right of people to remain in their country, the right of people to return to their country and the right of states to protect themselves from migrations that they perceive as upsetting international security.

Confronted with human rights violations and massive humanitarian emergencies following the breakdown of states, the international community claims increasingly that state sovereignty carries minimum responsibilities to maintain public order and protect human rights. When states, from volition or weakness, are not able to assure these minimum conditions, and humanitarian emergencies occur as a result, the international community has shown greater readiness to act by providing humanitarian relief and protection inside these countries. In several cases this has involved sending military contingents to protect aid deliveries or international representatives to monitor human rights conditions.

Spheres of influence

When individual governments see little national interest in distant crises, they are reluctant to take the lead in response and keen to see other governments take charge. This may explain the apparent reversion to thinking in terms of spheres of influence. By this logic, the former Yugoslavia was a European responsibility, Rwanda and Zaire were for France and Belgium to tackle, and the Caucasus and Central Asian republics are Russia's concern alone. It may also explain the reliance states have placed on the United Nations (UN) to take the lead in responding to humanitarian crises. An overstretched UN, on the other hand, increasingly seeks to devolve responsibilities and tasks to regional organisations such as the Organisation of African Unity, the North Atlantic Treaty Organisation, the European Community, etc. Finally, this reticence of governments to lead the response to distant humanitarian crises of little defined national interest may explain the increased reliance governments and intergovernmental organisations alike have placed on NGOs to advocate and lead the response to humanitarian crises.

With many changes in the political environment, significant adjustments – even experimentation – are occurring in refugee, and related humanitarian, policy and practice.

No longer propped up by Cold War interests, several long-standing regional conflicts came to an end, allowing voluntary repatriation by millions of refugees who had lived for years in camps. Peace accords were reached in Namibia, Ethiopia, Cambodia, Central America, Mozambique and elsewhere, allowing millions to return home. In the three years prior to the end of the Cold War, less than 300,000 refugees returned home annually; in the following three years voluntary repatriations averaged over two million per year. At the same time, new conflicts are producing an even higher number of uprooted people than that of returnees.

Significant challenges have had to be met to plan and implement programmes enabling refugees to return safely to their country. Once back, assistance was needed to help refugees, and the communities to which they were returning, cope with their reintegration. New forms of cooperation and coordination between relief and development agencies needed to evolve. UNHCR initiated what have become known as quick impact projects to aid communities receiving returning refugees. The UN Development Programme set up community-based, human development projects in Central America – known as PRODERE – and Cambodia – CARERE – to help the transition from relief to sustainable development. The success of these various efforts to help post-conflict societies is still being evaluated.

Not all Cold War conflicts were resolved when the Cold War ended. Some long-standing conflicts, such as Sudan and Sri Lanka, were not related to Cold War politics. Even where conflicts continue, pressures have increased for repatriation programmes to begin sooner rather than later. Some pressure comes from refugee communities. Community leaders of refugees returning to El Salvador and Guatemala insisted that repatriations begin before UNHCR thought conditions were suitable.

Amid rising domestic intolerance, countries of asylum have become more reluctant to host refugees, accurately assessing that while the international community will pay for emergencies, it is less willing to meet long-term costs.

During the 1980s, the war in Afghanistan produced four to five million refugees, the largest recent exodus. Donor countries funded major refugee assistance programmes in Pakistan. Even though UN efforts failed to halt the war after Soviet troops left in 1989, in the 1990s donor states began cutting their support. In response, UNHCR began providing aid packages to Afghan refugees wanting to repatriate. Between 1990 and 1995, this assistance and spontaneous return cut refugee numbers in Pakistan from approximately three million to perhaps one million.

This is not an isolated example of repatriation under less than ideal conditions. Recent repatriations – to Somalia from Kenya, Sri Lanka from India, Myanmar from Bangladesh and Rwanda from Zaire – show that donors and asylum countries are unwilling to support long-term refugee programmes and that, even under difficult circumstances, pressures to encourage repatriation will mount. The principle of voluntary repatriation

Box 1.3 How can environment create 'refugees'?

Environmental factors are increasingly cited as reasons why people are compelled to migrate internationally and internally. Similarly, there is concern about the impacts sudden, large-scale movements of people have on the environment. Estimates of the magnitude of what is often labelled the phenomenon of environmental refugees vary widely – and one is tempted to say wildly.

A World Bank study showed that public works projects now uproot more than ten million people a year in developing countries, four million of these ousted by the approximately 300 large dams constructed each year. It is estimated that urban development and the expansion of transportation infrastructure displaces a further six million people per annum.

In China, at present, three-quarters of its population lives in rural areas, but the Chinese Academy of Social Sciences believes that by the year 2010 half the country's population will be urban. This frightening trend has already caused concern. The Chinese government has started this year a clampdown on rural-urban migration and is displacing many people back to the land.

Global estimates of current levels of environmental refugees range from 10 million to 100 million and a projection (in a United Nations Environ-ment Programme report) is of over one billion environmental refugees by 2050. Such wide disparities in numbers indicate that there are major definitional problems with the concept of environmental refugees that need to be resolved if it is to have practical, operational applications.

Even in the broadest usage, the term "refugee" typically relates to situations where people are in jeopardy because of wars and conflicts and/or fear of human rights abuse. The term refugee implies a breakdown in the relationship between the individual and their state. In the vast majority of instances where people are forced to move because of environmental factors, there is not a fundamental break in the relationship between the individual and their state. In such cases it is not appropriate to use the refugee or even internal refugee label. This is not just a semantic issue. To make use of the term refugee to describe other types of problems can have the effect of eroding the protection rights afforded to real refugees.

If there are people fleeing across borders because of fear of violence or human rights abuses connected with development programmes or environmental catastrophes, then these people are refugees, period, regardless of the environmental context. ∎

is being stretched to include situations where refugees have little choice because conditions in the asylum country have become even worse than in their country of origin.

Following the Gulf war, the Iraqi army turned its wrath on Iraqi Kurds in Northern Iraq and Shiite Muslims along the southern border with Iran. Hundreds of thousands of people fled. Iran was sympathetic and provided the Shiites with a safe haven. When the Iraqi Kurds fled toward Turkey, the Turkish government blocked their entry. More than 300,000 Kurdish refugees became stranded in the mountains between Iraq and Turkey during the winter of 1990-1991. The international community was not prepared for the size and speed of this exodus and could not meet the needs in such difficult conditions. Under pressure from some governments, the UN General Assembly decided the Kurds had to return to their home areas. UN Security Council Resolution 688 declared that the exodus of the Kurds threatened international security, allowing the US military to create a safe zone in Northern Iraq, escort refugees back and then work with UN organisations to provide humanitarian relief programmes.

The Iraqi Kurdish emergency was the first major post-Cold War humanitarian crisis. The international response hinted at the constraints and opportunities for humanitarian action in future emergencies. Turkey's rejection of Iraqi Kurds was later reflected in the reticence of Western Europe to receive Bosnian Muslim refugees on other than a temporary basis. Perhaps this attitude towards mass refugee flows – in part on the grounds that such movements would disturb public order in receiving countries – was also present in the interdiction of Haitian and Cuban boat people by the US government. The response to the Iraqi Kurdish crisis signalled the intention of the UN General Assembly to address future emergencies in the countries of origin. The use of military capacities to create a safe zone in Northern Iraq laid the basis for operations in Somalia and the former Yugoslavia.

The Iraqi Kurd exodus and return shifted the vocabulary used to describe situations from "refugee crises" to "complex humanitarian emergencies", taking in refugees, internally displaced people and others still living in countries affected by conflict, human rights violations and other stresses. Until this time, humanitarian operations inside countries in conflict had been left, in the main, to the International Committee of the Red Cross (ICRC) with the major exceptions being the Office of Emergency Operations in Africa in Ethiopia in 1985 and, later, Operation Lifeline Sudan.

Figure 1.1 **Global figures: Refugees and IDPs**
There has been a steady rise in the number of refugees worldwide throughout the 1980s with a slight drop in numbers in the 1990s. Parallel to this, the numbers of internally displaced people continue to rise on a year-by-year basis. Predictions suggest that by the year 2000 we will see over 50 million people on the move, all needing assistance and protection and all competing for the same restricted humanitarian assistance.

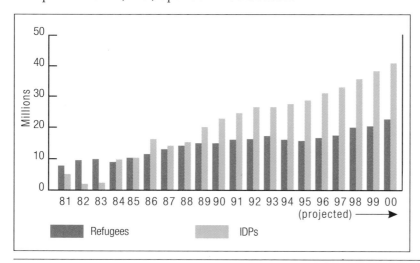

Source: US Committee for Refugees, Refugee Policy Group

With Iraq as a precedent, the UN system, operating under Security Council and General Assembly resolutions, has become more regularly involved in responding to the internal and the external dimensions of humanitarian emergencies caused by collapsed states, conflict and mass human rights violations. One aspect of this has been the expansion of UN peacekeeping operations to support both peace agreements and humanitarian operations amid continuing conflict. The creation of the UN Department of Humanitarian Affairs in 1991 to ensure more comprehensive UN responses to humanitarian emergencies was symptomatic of this shift.

Advocating action

Many reports on Somalia, Bosnia and Rwanda assert that the failure of the international community to take early and vigorous political action deepened the crises. Improved capabilities in emergency humanitarian relief do little good if they are deployed only after many people have died or if poor security prevents them from functioning effectively. Turning early warning into prompt response remains difficult, especially when the trigger of the crisis is political, rather than environmental or climatic, and donor governments' concern remains low unless their interests are directly affected or many lives are at risk. To overcome such political reticence, early warning information may be collected for advocacy rather than developed systematically for programme planning, which can lead to exaggeration of needs and thus inappropriate response.

Somalia, Bosnia and Rwanda reveal how important, yet difficult, it is to learn lessons from past emergencies, and demonstrate:
- that humanitarian action cannot be a substitute for – and may complicate – finding political solutions;
- that where massive needs exist, the donor-country media and public usually compel a humanitarian response;
- how hard it is to protect the rights and even the lives of internally displaced people exposed to conflicts;
- that humanitarians are exposed to greater risks in providing aid than in the past;
- that violence or its threat, through armed protection, complicates delivery of humanitarian relief; and
- how reluctant border countries are to provide secure asylum.

Response to Bosnia, Somalia and Rwanda revealed fundamental dilemmas. In Bosnia, agencies faced a delicate situation – trying to provide aid for people in "safe" havens in or near their homes often exposed them to attack. On the other hand, moving people made the humanitarian programme an instrument of ethnic cleansing. Another dilemma was whether relief saved lives and bought time for politicians to negotiate a solution, or merely gave politicians an excuse not to become more actively engaged in pursuing a solution to the political roots of the crisis.

In Somalia, relief supplies became targets for armed gangs associated with various factions. Sustaining relief operations needed increasing levels of security, including armed guards with their own factional affiliations. Many perceived the choice as being between the withdrawal of aid agencies or the arrival of external military forces to protect humanitarian supplies. Once the military arrived, their priorities took precedence over those of the humanitarian agencies whose operations they had come to protect.

The purpose of the military mission expanded – "mission creep" it was called, recalling US involvement in Viet Nam – beyond protection of humanitarian relief into the political sphere. The military had its own dilemma – it was not intended to stay long, but how could it leave if the security conditions it came to redress were not resolved? While the military presence helped temporarily restore the conditions allowing humanitarian operations and saved many lives, the forces finally pulled out with a sense of failure, most relief operations were scaled back significantly and no political settlement was reached.

The Somalia lesson that appears to have been transferred to Rwanda (see Chapter 7) was that of caution on the role external military forces can play in creating secure conditions. In Rwanda, half a million people, mostly Tutsis, were slaughtered by Hutus before the international community reacted. That reaction came only after several million Hutu refugees fled in the face of the advancing Tutsi army. Current Rwandan dilemmas include: How should the international community aid and protect Hutu refugees, many of whom were involved in the killings? Can the Hutu militia be prevented from recruiting in refugee camps for a reinvasion of Rwanda? Can the aid to refugees be balanced by aid to Rwanda's new, Tutsi-led government? How can the international community help and monitor the conditions of internally displaced people who have been encouraged to leave camps by the Rwandan government? Should the refugees be encouraged to repatriate to Rwanda from Zaire, Tanzania and Burundi? What aid and protection should be afforded by the international community to returnees, as well as to those refugees who will not be able to be reintegrated in Rwanda?

For 15 years after the end of the Viet Nam war, Vietnamese asylum seekers were automatically regarded as refugees. After first finding temporary asylum in refugee camps in South-East Asia, more than two million Vietnamese were resettled to the United States, Canada, France, Australia and other countries. By the end of the 1980s, those countries were far less willing to accept these refugees for resettlement and questioned the status of those leaving Viet Nam – were they refugees or economic migrants? Also, several states wanted to improve their political and economic relationships with Viet Nam. In 1989 the Comprehensive Plan of Action (CPA) was agreed between the Association of South-East Asian Nations, countries of resettlement and countries of origin. This allowed countries of temporary asylum, under UNHCR supervision, to decide whether individual new arrivals qualified as refugees.

Those judged to be refugees would be eligible for resettlement in third countries as before; those denied refugee status would remain in camps in the region until they returned to Viet Nam. Under the CPA some 70,000 "screened out" Vietnamese voluntarily went home, with the International Organization for Migration (IOM) arranging transport and UNHCR running

Figure 1.2 **UNHCR's funding: Changing funding profile**
Funding to the UNHCR has been increasingly dominated in recent years by Special Programmes, that is, emergency assistance and protection outside of the normal budget which is paid for from the General Programmes funds. Nearly 75 per cent of UNHCR's funding in 1995 went to Special Programmes, reflecting the shift in the agency's work from dealing with individual cases of refugee need and asylum seekers to the massive programmes of the former Yugoslavia and those for Rwandan refugees.

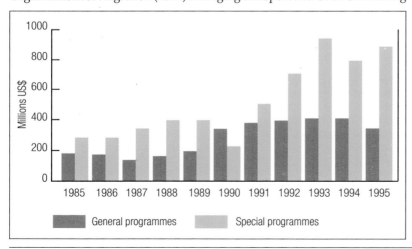

Source: UNHCR

monitoring and assistance programmes in Vietnamese areas receiving returnees. Some 40,000 Vietnamese denied refugee status continue to live in camps in the region. The CPA expires in mid-1996, raising questions over the fate of those remaining – will they be forced to go home, will UNHCR monitor and assist returnees in Viet Nam, or will special arrangements be made for resettlement outside Viet Nam?

UNHCR's mandate covers the monitoring of returning refugees but not those refused refugee status. UNHCR programmes to monitor and aid the return of non-refugees have been a departure from past practice. As in Viet Nam, the agency agreed in Sri Lanka to monitor the situation of Tamils refused asylum in Europe and deported back. In other instances, for example, Haitians stopped at sea and sent home by the United States, they declined to play a monitoring role. A German request for UNHCR to monitor the return of Vietnamese guest workers whose contracts had expired was refused. Similarly, IOM facilitated the return to countries of origin of guest workers expelled from Kuwait and other Gulf states at the time of the Gulf war. But in other situations where guest workers have been expelled, such as Ghanaians from Nigeria and Sudanese from Libya, no international organisation has played a role. In the future, it will remain a continuing challenge to sustain special understanding, mandates and protection for refugees while so many other compelling problems concerning population movement arise.

Restricted immigration

Both Europe and the United States are making efforts to restrict immigration, following years in which record levels of immigration took place, and the prospects for refugee resettlement or political asylum are worsening. For example, the US military occupied Haiti, under the terms of a UN resolution, to restore President Aristide to power rather than accommodate an emergency migration of Haitian boat people. The US also reversed three decades of policy granting Cubans asylum to reach an immigration agreement with the Cuban government. Potential changes in the US refugee resettlement programme could sharply cut the numbers allowed. Across Europe asylum procedures have been tightened, making them tougher and faster. Some European nations have reached agreements with states through which asylum seekers enter Europe to deport people back to these countries, without having to review asylum claims.

Despite significant East-West migration following the fall of the Berlin wall, fears of a more sustained migration were misplaced. Nevertheless, concerns remain about movements of people within the former Soviet bloc. Over 50 million people in the former Soviet Union live outside their titular republic. This figure includes over 25 million Russians living outside Russia in new states created along ethnic lines. Disadvantaged or unwelcome ethnic groups are on the move. The region's conflicts, from Armenia to Tajikistan, have all displaced substantial populations. Finally, countries in the region, particularly Russia, host migrants from countries such as Somalia and Iraq who claim to be refugees. Many may wish to migrate further west but Russia, having signed the Refugee Convention, must decide whether these individuals are refugees and, if so, accept them as such. These and other issues will be discussed in 1996 at an international conference sponsored by UNHCR in collaboration with the Conference on Security and Co-operation in Europe.

From the humanitarian emergencies of the past five years, a policy framework appears to be emerging. Three interrelated and fundamental shifts in approach appear to be in process:

● from emphasising the right of people to leave their country to emphasising the right to remain in their country and the importance of averting refugee flows and migration;

● from emphasising humanitarian assistance and protection in countries of asylum to emphasising humanitarian assistance and protection in countries of origin; and

● from emphasising the preservation of asylum to emphasising early solutions to refugee and displaced people situations, with priority on their return and reintegration.

These trends have implications for the development of international capacities and relationships between international agencies. Ideally, new practice should build from – not be a substitute for – established laws, principles and mandates. That cannot be guaranteed in the present policy climate, one less tolerant of migration, more determined to prevent or contain refugee flows and more impatient to find solutions, and to scale back international involvement.

Key donor governments are encouraging all parts of the UN system to pursue more comprehensive and coordinated approaches to humanitarian emergencies, believing that the advantages of this approach outweigh the disadvantages.

However, it will be harder for each organisation to implement its special mandate, and may impose new priorities or roles. UNHCR, for example, may press for earlier repatriation than it would if it were not being encouraged by key donor states to encourage such early repatriation. UNHCR also might mount assistance programmes for internally displaced people to prevent a refugee outflow. Both examples would make it more difficult for UNHCR to fulfil its protection responsibilities, in particular the right of people to seek asylum. The organisation may be encouraged – as in Viet Nam under the CPA – to monitor and provide assistance to non-refugees returning to their country of origin, and to act more regularly as the UN's principal operational agency in humanitarian emergencies, such as in Bosnia.

Outside the UN, NGOs and agencies such as the International Federation of Red Cross and Red Crescent Societies and the ICRC, whose strength has been based on independence, are increasingly urged to participate in comprehensive, coordinated humanitarian initiatives. In places like Somalia and Bosnia, UN organisations have provided relief amid conflict and used UN military units to protect relief deliveries. Such comprehensive approaches can affect the clarity of the roles taken by each part of the International Red Cross and Red Crescent Movement – ICRC, the International Federation and National Societies. This risks undermining the fundamental principles and laws within which they operate.

Fundamental change is affecting the causes and nature of refugee crises and the scope of response. Amid the transformed political climate since the end of the Cold War, the outlines are emerging of a policy framework within which refugee programmes will be implemented.

While humanitarian programmes should be impartial, they function within the prevailing political climate. The emerging policy framework contains both positive and negative components.

On the negative side, there are signs of the disengagement of key donor states from international aid programmes. This has not yet, but could in the future, led to constraints on humanitarian aid policies and resources. For this reason, humanitarian agencies cannot be complacent that recent large-scale increases in their budgets and programmes are sustainable.

There are positive signs, too, which need to be built upon. They include recognition that:

● international and regional capacities in both governmental and non-governmental organisations should be strengthened to prevent crises that uproot masses of people;

● emergency-response capacities should be sustained and strengthened;

Sources, references, further information

Bread for the World. *Countries in Crisis, Sixth Annual Report on the State of World Hunger.* Silver Spring, Maryland, 1995.

Cahill, Kevin, ed. *A Framework for Survival: Health, Human Rights and Humanitarian Assistance in Conflicts and Disasters.* New York: Council on Foreign Relations and Basic Books, 1993.

Cohen, Roberta. *Refugee and Internally Displaced Women: A Development Perspective.* Washington DC: The Refugee Policy Group and The Brookings Institution, 1996.

Deng, Francis. *Internally Displaced Persons: An Interim Report to the United Nations Secretary-General on Protection and Assistance.* Refugee Policy Group and the UN Department of Humanitarian Affairs, 1995.

Gallagher, Dennis. *The Evolution of the International Refugee System.* 1994. Refugee Policy Group, 1424 16th Street NW, Suite 401, Washington DC 20036, USA. Tel.: (1)(202) 387 3015; fax: (1)(202) 667 5034.

International Council of Voluntary Agencies. *The Reality of Aid '95.* Geneva, 1995.

Lawyers' Committee for Human Rights. *African Exodus: Refugee Crisis, Human Rights and the 1969 OAU Convention.* New York: The Lawyers' Committee for Human Rights, 1995.

Loescher, Gil. *Beyond Charity: International Cooperation and the Global Refugee Crisis.* Oxford: Oxford University Press, 1993.

Minear, Larry and Weiss, Thomas G. *Mercy Under Fire.* Boulder, Colorado: Westview Press, 1995.

Myers, Norman. *Environmental Exodus: An Emergent Crisis in the Global Arena.* Washington DC: The Climate Institute, 1995.

Refugee Policy Group. *Humanitarian Action in the Post Cold War Era.* Washington DC, 1992.

US Committee for Refugees. *World Refugee Survey.* Washington DC, 1995.

United Nations High Commissioner for Refugees. *The State of the World's Refugees: In Search of Solutions.* Geneva, 1995.

● response should be shaped within a broader framework based on improved awareness of the root causes, as well as the impact of humanitarian aid on the prospects for finding solutions;

● countries of origin have obligations to their citizens to protect the public order and basic human rights and that where their failure to provide these conditions results in the forcible displacement of people, the international community can act to aid and protect the victims;

● humanitarians should be prepared to cross borders to help people in their countries of origin rather than wait until a crisis has generated masses of refugees;

● the humanitarian community must work together in a cooperative manner to ensure that refugee and humanitarian problems are addressed efficiently and effectively;

● once crises are resolved, international efforts are required to keep the crisis from re-emerging; and

● despite cutbacks for international programmes generally, donor states and private citizens appear to be willing to continue to invest in these efforts.

The challenge to the community of organisations concerned with refugee and related humanitarian problems is to bring out the best of the current political and policy context. In this changing climate, each crisis and its response is a laboratory that provides humanitarian organisations the opportunity to learn from their successes and failures.

In the future there will be no shortage of refugee and related humanitarian crises, no doubt in new and confounding forms. The best hope for meeting these challenges is in preserving and strengthening capacities that have proven to work, abandoning those which are no longer relevant and adapting capacities to new circumstances and times. ∎

Figure 1.3 **Returnee populations: Returning to what?**
The ending of the Cold War has seen some positive effects on refugee populations. In Afghanistan and Mozambique, millions of people have returned home, but they have returned to countries ravaged by war and, in the case of Afghanistan, still ravaged by internal fighting. For many of these returning populations, and the host populations who receive them, the need for external assistance is still present. Many cannot return to their original homes and end up ebbing and flowing across the country as fighting flares. Former refugees often become today's internally displaced.

Source: US Committee for Refugees

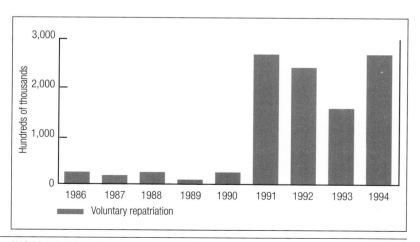

Hearts, minds, stomachs: Along with water, shelter and medical care, food is a key need for those caught up in disasters. A tight cereal supply in early 1996 pushed up prices and cut stocks well below levels the FAO believes necessary for global food security. Greater difficulties in food-aid costs and supply came as poor countries faced higher import prices. As well as feeding people, food aid can be a political tool. In many complex emergencies, food aid may not be used for purely humanitarian purposes.

Soldier with food aid, Chiapas, Mexico, 1995. Larry Towell/Magnum

Food enough for the needy?

Images of hunger are all around us. Pained faces of children staring out from the war-zone refugee camps of sub-Saharan Africa. Slight Bangladeshi women struggling to survive in a rural environment that values them less – and feeds them less – than their male kin. Men searching for food in the impoverished inner cities of an otherwise wealthy United States.

This in 1996, a year in which the world will again produce more than enough food to provide every human being with an adequate diet. The persistence of hunger shames the world. Despite 50 years of United Nations-led human rights declarations that all people deserve a decent standard of living, good nutrition remains beyond the reach of three-quarters of a billion people.

This chapter will examine the reasons – ecological, political, economic, cultural – for such widespread misery, and consider why millions lack economic access to land, resources and livelihood, and millions more suffer outright cultural discrimination and denial of healthy diets.

Data assembled by United Nations (UN) agencies, especially the Subcommittee on Nutrition of the UN Administrative Committee on Coordination, and the Brown University World Hunger Program indicate that:
- more than three-quarters of a billion people lack sufficient food energy to live fully productive lives;
- 190 million pre-school children risk vitamin A deficiency, iodine deficiency menaces more than a billion people, and approximately 40 per cent of the world's population – including 50 per cent of all women and 79 per cent of all pregnant women – suffer iron-deficiency anaemia; and
- at least 184 million children under five are underweight for their age.

By world regions, figures for the malnourished are highest in Asia but numbers are growing most rapidly in Africa, and Latin America continues to suffer endemic poverty-related malnutrition. These statistics include individuals in wealthy countries, such as the United States, where an estimated 30 million – one in nine Americans – go hungry due to poverty at some point each month.

These numbers persist alongside findings that:
- since the 1970s the world has produced enough food to provide 10 to 20 per cent more than its population with an adequate per capita calorie intake of 2,350 kcal/day;

- absolute and relative numbers of malnourished people have fallen since 1975. Trends towards adequate nutrition in Asia more than offset deterioration in Latin America and sub-Saharan Africa;
- technical and institutional capacities to eliminate malnutrition and hunger-related illness due to micronutrient deficiencies are growing, especially in the developing world; and
- early warning systems to monitor and respond to conditions of acute food insecurity are eliminating most famine deaths in traditionally famine-prone countries, except in war zones.

Thus some progress has been possible. There are the beginnings of an international political will to affirm adequate food and nutrition as a human right.

Hunger is often equated with famine – an acute food shortage leading to widespread deaths and movements of people searching for sustenance. But insufficient nutrient intake and food insecurity – individuals or households

Box 2.1 Food secure or insecure?

Throughout the 1970s and early 1980s world food production rose at an average 3 per cent a year. Improved seeds and fertilisers, and the cultivation of previously marginal land accounted for most of this increase. There is enough food in the world today to give everyone an adequate diet, if it were evenly distributed. It is not, and there is much disagreement about whether past trends in production and yields will carry on into the future.

Red Cross and Red Crescent Societies around the world are regularly involved in addressing both acute hunger in emergencies and chronic hunger in longer-term health programmes. Food poverty (security) and deprivation (malnutrition) will continue to be challenges that the International Federation and other agencies must address if they are help peoples of the world achieve their right to food and freedom from hunger. Erring on the side of caution, the Federation believes it is prudent to pay heed to the "concerned" case; the "confident" case will only become a reality if we take action now to ensure the research translates into reality and policy into practice.

The 'confident' case

This case, championed by the the Food and Agriculture Organization (FAO), hinges on the belief that sustained agricultural research and the translation of that research into higher yielding crops and more efficient production, will enable world agriculture to keep on growing at a rate just ahead of population growth.

FAO believes that the growth rate of agricultural production in the next 20 years will be slightly less than in the past. When this is combined with best predictions for population growth, it will result in a slight falling-off of world per capita cereal production from 327 kg in 1989-1990 to maybe 325 kg in 2010. However, within this global figure, cereal production in developing countries is forecast to rise. FAO predicts that the present average calorie availability in developing countries (2,500 per day) should increase to 2,700 by the year 2010, with China, East Asia and Latin America reaching 3,000 calories per day. The prospects for Africa, though, are not so good.

The 'concerned' case

The Washington-based Worldwatch Institute, the most well-known and vocal champion of the concerned case, points out that there is very little marginal land left to bring into production and that since 1984 the annual rate of food-production increase has slowed down to nearer 1 per cent. Since 1974, almost all the increase in global food production has come from yield increases, not from extra land, and total cereal output per capita has been in decline since 1982. Throughout the 1980s food production failed to keep pace with population growth in 75 developing countries.

Beyond the immediate and recent-past factors affecting grain production, there are several longer-term global trends that many believe are leading to a loss of momentum in the growth of world grain production. One is the scarcity of new land that can be brought under the plough.

Another is the widening scarcity of usable water for irrigation. And finally there is the diminishing response of crop yields to the use of additional fertiliser. ∎

not knowing how to meet their minimal needs – is much more a problem of economic access to the right nutrients needed to lead a full and healthy life.

Three distinct but related concepts are used to estimate the numbers of people affected by hunger and to analyse the global food situation: food shortage, seen at its most extreme in famine; food poverty or lack of access; and food deprivation leading to malnutrition.

The hunger typology frames the levels of suffering, suggests opportunities for intervention, raises a basic question of hunger and human rights – who is left out of the community's obligation to assure everyone adequate food? – and suggests who might take responsibility for preventing such hunger.

This framework also highlights the impact of disasters, especially war, on food security. Armed conflict remains a principal cause of hunger, and food insecurity is itself a factor provoking conflict. In 1995, at least 32 countries saw armed conflict in which hunger was an integral part of the conflict, and hunger problems in at least ten more countries were due in great part to the legacy of war.

Food shortage

Food shortage occurs where total supplies in an area from production, and imports minus exports, are insufficient to meet the needs of the area's population due to political, climatic or socio-economic forces. Food production data from the Food and Agriculture Organization (FAO) show that, in

---

Box 2.2 Mexico – food, land and rebellion

Mexico's Zapatista revolution erupted on 1 January 1994, as economically impoverished and politically powerless communities of ethnic Maya demanded food, land and livelihoods.

In response, the army allegedly brutalised civilian populations. Houses were looted, food stocks and livestock destroyed in operations that used hunger as a weapon against possible insurgents. Tens of thousands fled into the forest, where they had no access to cultivable land or work.

These people became entirely dependent on the humanitarian handouts permitted by the military, who widely publicised their own food distribution, vaccination, dental and medical care. There were also reports of looting by Zapatistas, who allegedly picked the coffee crop of some of those who had fled, thereby subjecting those displaced to penury when they returned.

The conflict raged on into 1995, with new army attacks reportedly devastating households, contaminating water sources, and driving pro-Zapatistas and whole villages into the mountains. Although some displaced were being offered economic incentives to return (if they reported on rebel activities), over 15,000 were said to be displaced in the hills without means of subsistence.

The conflict began on the eve of the new North American Free Trade Agreement, perceived to threaten indigenous livelihoods already undermined by cattle farming and other predatory economic interests. Government poverty-alleviation programmes – *Solidaridad* – have been largely ineffective at removing the sources of structural violence or improving livelihoods. Indigenous populations still lack access to adequate land, transport, education or health services.

The roots of rebellion can be traced to generations of inequality, aided and abetted by indigenous leaders who conspired with the Mexican elite to maintain control over land and natural resources. Rebellion against human rights abuses and inequality came in the context of more than a generation of exposure to the ideas of Catholic Action (liberation theology), Marxism and the economic influences of non-governmental organisations.

In early 1996, the Zapatistas, after extensive negotiations with the government, are trying to work through the political process to address their demands. Meanwhile, amid more widespread political violence in the aftermath of the Chiapas rebellion, Mexico is struggling to control soaring inflation and economic disorder, itself likely to lead to additional conflict as unemployment and price rises further reduce access to food, causing malnutrition. ∎

1995, the world as a whole suffered no food shortage. But FAO data also show that, after subtracting losses to pests, in storage, and to conversion of foods into animal protein or industrial uses, 48 states with a total population of 802 million had less dietary energy supply than expected needs.

The data also show that acute shortages, or famines, are almost always conflict-related, although nutritional shortages also follow in the wake of rapid economic and political changes, such as those from central command to market economies.

Conflict creates food shortage by the deliberate use of food as a weapon. Adversaries create food scarcities to starve opponents into submission in many ways: seizing or ruining food stocks, cutting off food sources and destroying markets, diverting relief food from beneficiaries, or contaminating land and water resources. Anti-personnel mines are another factor in many countries, ensuring long-term underproduction and vulnerability to food shortages. In some countries as much as 35 per cent of arable land cannot be used because of mines. Food shortages also occur as people flee or fear to farm agricultural lands.

In sub-Saharan Africa, extreme food shortage was a key factor in the conflicts of Angola, Southern Sudan and Somalia and food relief was manipulated as a political weapon by warring parties. Rwanda faced severe food shortages as violence drove millions from their homes. Farmers were unable to return to plant, setting the stage for multi-year food shortages and reliance on outside aid. In Kenya, a political-economic struggle for land between herders and farmers has caused food shortages and driven tens of thousands from their land.

In Asia, Cambodia was subject to extreme food shortage, while siege warfare, hunger and struggles for control of relief food characterise continuing conflicts from the former Soviet Union and Yugoslavia to Afghanistan, Myanmar and Sri Lanka. Local populations caught up in the conflicts of Mexico and Guatemala were also short of food; denied access to land or livelihood, they were dependent on the military for relief. In each case, destructive arms and human displacements compound problems of land and livelihood shortages that perpetuate hunger and conflict.

Food poverty   Entitlement (sen)

Food poverty refers to the situation in which households cannot obtain enough food to meet the needs of all their members, even if there is ample food available in the market or stores. The 1992 International Conference on Nutrition reported a decline in the food-poor in the 1970s and 1980s, with an absolute decline in numbers between 1975 and 1990 from 976 million to 786 million.

The positive trend masks regional differences. In the 1980s and 1990s in sub-Saharan Africa and South America both the proportion and number of food-poor have increased. Trends are also negative in the former Eastern bloc, where economic changes have caused an alarming deterioration in health and nutrition. But positive trends reported by Asian governments, especially China, more than outweigh declines elsewhere.

Principal sources of food poverty are lack of access to land, to livelihoods, and to safety-net programmes that might otherwise meet food gaps. A special concern is livelihood failure due to warfare, both to those caught in active- or post-conflict zones that have been destroyed by warfare, and also refugees and displaced people who lose their land and sources of livelihood overnight. Growing urban poverty is another concern, in both developed and developing countries, while at the same time rural-urban migration continues.

Less evident than food shortages, livelihood and entitlement failures are also linked to conflict. In post-war Eritrea, for example, people recreating their communities must first gain titles to land, rebuild waterworks, replant trees for fuel, food and fodder, and get access to seeds and animals. From Mozambique to Central America, demobilised soldiers who have known only war need to acquire land, tools and skills to avert future food insecurity.

As wars cease and countries aim for social reconstruction and food security, food aid provides immediate relief but other strategies are needed to rebuild local capacities for food security and self-reliance. "Food-for-work" schemes to create employment and infrastructure and "linking relief to development" (LRD) strategies are expensive and involve trade-offs. Jobs may be created at the expense of feeding the most hungry. In countries with weak central governments, such as Ethiopia, LRD strategies suffer from a low capacity to plan and implement such programmes. Working with local communities may be difficult if many people have been in exile for years.

Box 2.3 Food security – the regional dimension

While predicting global food security may be difficult, the future seems fairly clear for some regions of the world, for others – like China – it is certain that their future food security will have a major influence on the situation for the rest of the world.

China

China imported 3.3 million tonnes of grain in the first half of 1994; in the first six months of 1995, 6.4 million tonnes were imported. Many experts predict that China's growing population will, within 35 years, have outstripped not only its ability to produce food, but will also absorb all the surplus grain production of the United States and Europe.

Countering these pessimistic arguments, Chinese officials point to recent land surveys using new satellite techniques which have revealed that China has some 38 per cent more cultivable land than previously accounted.

Whether the optimists or the pessimists are correct, food security in China, the country which still "boasts" the world's most devastating historical famines, represents one of the largest unknown factors shaping global food security

Former Soviet Union

Under its centralist economy, the former Soviet Union (FSU) had one of the world's highest per capita cereal consumption rates. This was because of high subsidies on consumer food prices and animal feeds.

Most predictions assume that the FSU will turn from being a net importer of grain (31 million tonnes in 1991) to a net exporter, perhaps exporting as much as 15 million tonnes annually by 2020. It is assumed that this will be achieved through lowering in-country grain consumption and through improvements in harvesting, storage and grain transport techniques. No increase in land under cultivation or grain yields is expected.

This scenario, however, assumes that the countries of the FSU are able to make the transition from centralist to market economies without mass violence. If Russia and the northern FSU states go the way of the Caucasus, then the FSU is likely to be a net importer of grain in the long term.

Greater Horn of Africa

In the Greater Horn of Africa, acute and chronic food insecurity is more severe than in any other region in sub-Saharan Africa, and is increasing because of a complex and interrelated set of political, social and economic factors.

In 1989, an estimated 46 per cent of the region's population, some 71 million people, were chronically food insecure. This percentage is greater than that of the overall figure for sub-Saharan Africa. In 1994, approximately 22 million people required external food-aid assistance. Nearly 11 million of these were refugees and internally displaced people, with another 11 million in danger of being severely drought-affected. Per capita food production declined in the region by more than 16 per cent during the period from 1980 to 1993. As a result, domestic food production per person has declined and food-import bills have placed increasing strains on trade balances. Donor food-aid assistance has also increased.

Daily calorie availability per person in the region (1,950 kcals) is less than the international minimum standard for survival of 2,100 kcals, and much less than the standard for an adequate diet of 2,400 kcals a day. ∎

Some economists have long advocated a "trickle-down" model of development, insisting that economic growth eliminates poverty and hunger. They highlight Asian nations that report cutting poverty and hunger substantially over a 15-year period characterised by growth. However, most policy planners today recognise that growth is a necessary, but not a sufficient, condition to eliminate poverty-related hunger.

Food security is inextricably tied to respect for basic human rights and implementation of poverty-reduction and social-welfare programmes. At a political level, a free press and vigorous community participation in government are essential. The hungry must have a voice and channels through which to be heard. Government regulations on minimum wages, land rights, access to communications and transport, fair-priced food, and safety-net social security programmes all favour improved incomes, nutrition and health. Government programmes in education, health and sanitation also are critical for improving food security. Well-targeted food subsidies, food coupon programmes and labour-intensive public works employment schemes help fill the food gap caused by underemployment.

Social service and social security programmes are more characteristic of urban than rural areas. More government investments in rural areas might help reduce incentives for the rural poor to move to cities, and decrease both rural and urban poverty-related hunger. Government action to improve women's health, employment and training should also increase women's capacities to contribute to food security.

Food deprivation

Food deprivation refers to malnutrition: the inadequate individual consumption of food or of specific nutrients. It can be due to cultural rules denying people sufficient access, or health conditions rendering them unable to ingest, metabolise and benefit from nutrients potentially available. Discrimination and deprivation may be based on age, gender, or other social classifications, including ethnic, geographic, religious, occupational or "outsider" – such as refugee – status.

Africa	Asia	Latin America	Europe
Algeria*	Afghanistan*	Colombia*	Armenia*
Angola*	Cambodia*	El Salvador	Azerbaijan*
Burundi*	India (Kashmir)*	Guatemala*	Bosnia-Herzegovina*
Eritrea	Indonesia (East Timor/West Irian)*	Haiti	Croatia*
Ethiopia		Mexico	Chechnya (Russia)*
Ghana*	Iraq*	Nicaragua	Georgia*
Kenya*	Myanmar	Peru	Moldova*
Liberia*	Philippines		Serbia*
Mozambique	Sri Lanka*		Tajikistan
Niger*	Turkey*		
Nigeria*			
Rwanda*			
Sierra Leone*			
Somalia*			
Sudan*			
Togo*			
Uganda			
Zaire*			

*denotes cases of active conflict where hunger has been used as a weapon.

Figure 2.1 **Food wars: Countries affected by food wars**
Food, its control and its destruction, is a prime weapon used in war. Land is rendered unusable, household food stores are destroyed, cities are blockaded and crops are requisitioned for armies, yet international humanitarian law prohibits the use of starvation of civilians as a method of warfare and the attacking, destroying, removing or rendering useless, for that purpose, of objects indispensable to the survival of the civilian population.

Source: Brown University, World Hunger Program

Deprivation also denotes imbalanced nutrition, "hidden hunger" which affects mainly women and children, and whose impact on mortality is masked by disease, such as micronutrient deficiencies, especially of vitamin A, iodine and iron, all of which in their extreme forms cause disorders themselves and in moderation are factors detracting from health and human functioning.

More than one billion people are liable to suffer iodine deficiency disorder (IDD) that each year permanently blocks the intellectual and motor development of one million children and retards the mental and physical growth of five million. For five US cents per person per year, salt iodisation can prevent iodine deficiency. Since the 1990 World Summit for Children, which set 95 per cent iodisation rates as a mid-decade goal, 58 countries with almost 60 per cent of the developing world's children are close to achieving that goal, and another 32 countries might achieve it with extra effort. The former Soviet Union and Eastern Europe, however, show rising levels of IDD.

Vitamin A deficiency at its worst causes permanent blindness in children and, in a less extreme form, night blindness and increased morbidity and mortality, especially from measles. Approximately 13.8 million children suffer the severe symptoms of this deficiency and another 176 million are at risk. Since the World Summit for Children, there has been marked progress in eliminating this micronutrient deficiency, through a combination of capsule distributions to under-fives in health clinics, fortification of sugar, and nutrition education to encourage consumption of vitamin A-rich foods. Approximately two-thirds of the children at risk of this deficiency live in countries where governments are pursuing such efforts.

Iron deficiency potentially incapacitates up to 40 per cent of the world's population, especially women and children. It is relatively inexpensive to prevent, despite the daunting logistics of getting people to take supplements. Progress reported from China, where weekly rather than daily supplements proved sufficient, may improve compliance and cut costs.

In recent years, the links between mild-to-moderate deficiency diseases and human functioning have been clearly demonstrated. One eminent Cornell University nutritionist has tried to quantify protein-energy malnutrition's contribution to child mortality. Although infectious disease is the main cause of child mortality in the developing world, his studies show that malnutrition – predominantly its mild-to-moderate form – contributed to 56 per cent of all child deaths.

Women and children at risk

Women and children are especially vulnerable to food deprivation, particularly when, in warfare, they are left behind as men migrate to find food or work, or are conscripted into armies. They also suffer disproportionately from disease, such as respiratory and gastrointestinal disorders or cholera, as – even where food is present – a combination of malnutrition and the lack of sanitation, water and health services renders them more at risk.

Reporting in 1993 on nutritionally vulnerable children, the UN Children's Fund (UNICEF) estimated that, in a decade, conflict killed more than 1.5 million, left at least four million physically disabled and made more than 12 million homeless. Emergency food rations, supplying calories predominantly through cereals, legumes and edible oils, often leave children especially vulnerable to micronutrient and other nutritional deficiencies.

Food deprivation also pinpoints who is left out in the division of food within households. The Indian economist Amartya Sen has documented how females face particular discrimination, as seen in statistics of their lower survival rates in many cultures, such as China and South Asia.

The female-to-male ratio at every age should be close to 1 (actually just over 1, given the survival advantages of women), but by the mid-1980s in China it was 0.941, in Bangladesh 0.940, India 0.933, and Pakistan 0.905. Tens of millions of women are "missing" in excess mortality due to a lower access to food and health care: neglected relative to their fathers, husbands, brothers, they also face male violence, especially amid extreme poverty. Another concern is targeting reproductive-age women to receive adequate micronutrients, especially iodine.

Women are often the victims of violence and hunger, especially in refugee situations, such as Rwanda. Critics of existing food distribution systems, such as the non-governmental organisation (NGO), African Rights, urge direct distribution to women to ensure more adequate nutrition for them and their children, rather than allow male diversion of food.

International sanctions are also a source of nutritional deprivation, although their immediate impact is via food shortage and poverty. Essential foods and medicines are explicitly excluded from such embargoes, but the poor are affected most as prices rise and supplies fall through transport problems and energy shortages, while a declining economy reduces their buying power. For a fuller discussion of these issues, see Chapter 2 of the *World Disasters Report 1995* (UN sanctions and the humanitarian crisis).

Food shortage, food poverty and food deprivation are interrelated. Too little food within a region means some households have insufficient access, and households without adequate access deprive certain members of food. Such hunger has common underlying factors, especially conflict and international economic policies that fail to put people and food security first in their priorities.

Conflict conundrum

Conflict disrupts food production, marketing, livelihoods, and nutrition and health care. Food insecurity also contributes to conflict. This hunger-conflict-hunger conundrum is a persistent source of food insecurity. The impact of peace was calculated by a Brown University World Hunger Program project – had sub-Saharan African countries, mired in conflict for 20 years, been free of war, they might have added 2 to 3 per cent to regional production per capita per year.

Peace and good governance also differentiate countries in their response to food crises. "Failed states", such as Somalia, have limited capacity to respond, even with outside help, to a food crisis. In the Southern African drought of the early 1990s, Zimbabwe, Botswana and other nations that successfully confronted drought-related food shortfalls were the region's most stable states; their governments had early warning and response systems and, with donor assistance, staved off food and political crises, famine, and conflict.

Food shortages and poverty – especially rural poverty and malnutrition in the developing world – are also blamed on international economic policies that discourage local food production for local consumption.

Figure 2.2 **Who gets to eat? Different ways of sharing the world's food basket**
As Gandhi said, "the earth has sufficient for every man's need, but not for every man's greed". There is sufficient food in the world today to feed all the globe's population, but access to that food and entitlement to it through production, purchasing or custom is not universally enjoyed.

Source: FAO, 1994 SOFA database

Level of diet	How far would present world food supply go on this diet?
Basic diet	6.26 billion (112 per cent of world population)
Improved diet	4.12 billion (74 per cent of world population)
Full-but-healthy diet	3.16 billion (56 per cent of world population)

Instead of putting "food first", developing countries have been urged to promote cash crops in which they are supposed to have a comparative advantage. Unfortunately, farmers who follow this advice often find themselves short of both cash and food. Adequate food and "development", "food first" advocates argue, is not a question of developing the right technologies or getting international grain prices right, but one of entitlements, social justice and empowering the poor. If poor people had access to resources, from land to education, there would be no food problem. The hunger problem, they conclude, is not one of too little food or too many people, but unfair distribution. The relationship between food production and hunger, however, is complex.

From 1986 to 1988, 99 countries imported more food than they exported, suggesting a high level of interdependence among food-trade nations, although not necessarily hunger in any particular one. Were there no food trade or aid, there would be more, not less hunger. Countries that are not food self-sufficient, however, are subject to price fluctuations for both food imports and crop exports, leaving earnings beyond their control.

Trade liberalisation and economic "adjustment" policies complicate the picture. Predominantly agricultural countries that promote cash crops to generate foreign exchange may find themselves facing higher prices for food imports. Very poor countries that are highly reliant on donated food may find themselves unable to make up shortfalls as available food aid shrinks.

For farmers and agricultural labourers who rely on food production for nutrition and income, declining food self-sufficiency also may reflect declining entitlement to land and other resources, and their increasing vulnerability to market fluctuations for their main cash crops, as happened in Rwanda. Nevertheless, most studies show that the best strategy for farmers in the developing world has been to diversify production and livelihood strategies, with cash crops and livestock alongside some subsistence production. This assures them some food if cash crops fail or food prices soar. The challenge is to create food interdependence rather than dependence.

Future food

Future food security projections vary widely. Some, such as ecologist Paul Ehrlich, insist that a food catastrophe already exists. Lester Brown and colleagues at the Worldwatch Institute foresee disaster soon, given population growth, limits on technological progress, and environmental decline. Per Pinstrup Anderson and Rajul Pandya-Lorch of the International Food Policy Research Institute (IFPRI) concur that a food crisis is underway in some

Figure 2.3 **Hungry populations: Populations with insufficient daily energy supply**
In every continent of the world there are countries that cannot feed their entire populations with the food reported to be available. Within their borders, some go hungry (although even in calorie-deficient countries, some eat enough or too much, reducing the amounts available for others). Sub-Saharan Africa hosts the largest number of countries with lower than adequate dietary-energy supplies. It is in this region that future famines are also most likely.

Source: UNDP and UNICEF

Region	Population in millions
Sub-Saharan Africa	459.1
Near East and North Africa	12.5
Asia	262.4
Latin America	67.2
North America, Australia, Western and Eastern Europe, the former Soviet Union, and small islands	1.1
Total	802.3

regions, such as sub-Saharan Africa, where per capita food supplies are falling, and in South and East Asia, where yields may be levelling off. They hope investment in agricultural research can reverse food security declines.

Some are more optimistic. Some World Bank economists insist that continued investment in agricultural research and technology will pay off in increased agricultural production, and there will be no food production problem for the foreseeable future. An NGO project, looking towards 2050, calculated how much more food might become available by greater savings in production, distribution and consumption. Efficiencies are possible even without greater overall production, for example, by growing more moisture-sparing crops in drought zones, cutting storage wastes, and changing consumer behaviour to eat lower on the food chain, i.e., more grain, less meat.

Alex McCalla, chair of the technological advisory group of the Consultative Group on International Agricultural Research, that sets priorities and funding for research centres, reviewed the picture as mixed. For 30 years the centres substantially raised yields in basic food crops – wheat, rice and potatoes – and to a lesser extent in other grains, tubers and vegetables. McCalla says there is little disagreement on future food needs – increases of 2 per cent or more per year – but arguments persist about where, if at all, the food will come from.

World Bank economists assume that developing countries will be able to sustain 3 per cent increases in food production per year. They do not anticipate soil, water, seed, chemical or price constraints on such progress. IFPRI economists find such an optimistic scenario likely only with immediate substantial investments in agriculture. Worldwatch Institute pessimists are not sure any increases will come, given environmental degradation, water scarcity and the impact of intensive agriculture.

This last view challenges another scenario, in which developed agricultural countries expand production and make up the "food gap" of developing countries, which would increase their food imports and industrial exports. Most economists question whether such intensified production is environmentally sustainable, whether producers and consumers, food donors and recipients, could agree on prices, and whether the infrastructure could handle a four-fold increase in food trade.

Food scenarios

Future food scenarios must account for a growing demand for higher-cost cereals, such as rice and wheat over sorghum or root crops, and more resource-expensive animal products. Livestock conversion is usually calculated in terms of food energy grain-to-livestock ratios. In a feedlot it takes two kilos of grain to produce one kilo of chicken or fish; a 4:1 ratio for pork and 7:1 for beef. Some suggest higher figures, at 3:1, 6:1 and 16:1. Even now, 4.3 billion large domesticated animals and 17 billion poultry eat 40 per cent of the world's grain supply.

Animal production also takes land and water resources that might be otherwise allocated to less resource-expensive food crops, but the livestock economy is complex, and reductions in animal production will not automatically produce more food at lower cost for the poor. Such shifts in where people live and what they choose to eat raise questions about government policies designed to assure food production and distribution for all these people.

Other factors affecting future food scenarios are the costs and controls on factors of production, especially water, while the food-population equation depends on assumptions about limiting population growth through economic and educational incentives, access to family planning and confidence that children will survive.

Whatever the future holds for the world balance of food production and consumption, one thing is clear, food aid – both to address chronic needs and, more critically, emergency needs – will remain a key last resort to fight hunger.

Food aid is essentially of three different types: programme, project and emergency. There is a clear trend for developmental food aid to decline as emergency use climbs, while the amount of food aid available is not keeping pace with food-aid needs.

Programme food aid: Provided as a grant or soft loan on a government-to-government basis, it helps to fill the gap between demand at existing income levels and the supply of food from domestic production and commercial imports. It relieves pressure on food prices, cuts demands for imports and provides balance-of-payments support by reducing the need to spend hard currency on food imports.

Project food aid: Aimed at transferring income to the poor or meeting their minimal nutritional needs in normal years. The food is provided – usually by the WFP – on a grant basis for specific targeted populations and to meet specific developmental needs.

Emergency food aid: Provided in response to sudden disasters, war or other calamities. Almost all is provided free and targeted to specific populations as a direct nutritional support. Much of it now comes through the World Food Programme (WFP) and much of it is distributed via the International Red Cross and Red Crescent Movement and NGOs.

For 20 years the world has used 6 to 13 million tonnes of food aid a year, most of it cereals. Food aid represents both a declining portion of global development assistance and agricultural trade. In the early 1990s, annual food aid from all donors totalled $3.4 billion, representing only about 6 per cent of total overseas development assistance, down from 12 to 15 per cent in the mid-1970s.

Several factors contributed to this relative decline, including reduced availability of US "surplus" food commodities and donor preferences for more flexible development assistance. Cereal food aid, which accounts for more than 90 per cent of world food aid, amounted to only about 5 per cent of global trade in cereals in the early 1990s, compared to nearly 10 per cent in the early 1970s.

Signatories to the 1986 Food Aid Convention – Argentina, Australia, Austria, Canada, the European Community, Finland, Japan, Norway, Sweden, Switzerland and the United States – guaranteed to provide a minimum of 7.5 million tonnes of cereal food aid each year, most of this through WFP.

Figure 2.4 **Micronutrient malnutrition: Numbers of people, in millions, affected by micronutrient malnutrition**

It is not just total calorific intake that determines good nutrition. For substantial populations around the world, deficiencies in iodine, vitamin A or iron cause poor diets, which in turn cause poor health, a reduced ability to work and hence a reduced ability to acquire a good diet. The vicious cycle must be broken.

Source: WHO, 1991, 1993

| Region | Iodine | | Vitamin A (pre-school children) | | Iron |
	At risk	Goiter	At risk	Xerophthalmia	Anaemia
Africa	150	39	18	1.3	206
Asia & Oceania	685	130	157	11.4	1,674
Americas	55	30	2	0.1	94
Europe	82	14	0	0.0	27
Eastern Mediterranean	33	12	13	1.0	149
World	1,005	225	190	13.8	2,150

The 1995 convention renegotiations cut the amount guaranteed by over 2 million tonnes. The convention underpins the global emergency food aid but it does not paint the whole picture.

In the 1980s and 1990s, two key trends in food aid emerged. First, the flow of food aid, particularly emergency food aid, has been consolidated through WFP. It now provides food for 57 million people, of which 32 million receive emergency food aid and 25 million project and programme food aid. Second, when WFP was created in the 1960s, it was primarily an agency delivering project food aid. Now 59 per cent of food deliveries are for emergencies and refugees, the rest for development or bilateral projects. WFP has become an emergency food aid organisation.

Food-aid trends are difficult to forecast. One recent study predicts that food-aid donations in 1996 will be around 9.2 million tonnes, reflecting the reduced convention pledges, and that donations will only rise marginally in ten years to reach 10.6 million tonnes by 2005.

The study's best-case scenario predicts that total world food aid needs will increase to 39.79 million tonnes, an increase of 5.29 million tonnes on the most optimistic predictions for 1996's needs. Worst-case scenarios put the global needs at 48.32 million tonnes and all this is set against a predicted rise in food-aid availability of less than 2 million tonnes.

Like all such predictions, it provides useful guidance as to where we need to watch for changes, but it is also important to understand the assumptions behind such predictions if one is to use them correctly.

The prediction that food-aid availability will only rise to 10.6 million tonnes in 2005 is based on a number of research assumptions:
● that the combined food-aid budgets of all donors will remain constant, after adjusting for inflation, at 1994/1995 levels but that the volume of grain that can be purchased with these budgets will rise because of slowly falling real prices of grain;
● the scenario also assumes that the ability of donor countries to produce sufficient grain to meet global demand from commercial importers will not be a constraint in 2005; and
● commercial demand from China and a few other importers is expected to rise during the projection period. If China seeks to increase its grain imports significantly, as some scenarios suggest, this will increase world grain prices and stifle consumption, stimulate production and constrain use of scarce foreign exchange for imports.

Chronic needs arise from a country's sustained inability to produce enough food for its population or to earn enough foreign exchange to commercially import the balance needed. Chronic needs are generally related to a state's long-term structural, resource or policy problems.

Emergency food-aid needs are caused by short-term production shortfalls due to weather problems, sudden natural disasters or political instability.

Sources, references, further information

Bender, William. "An End Use Analysis of Global Food Requirements," *Food Policy*, 19:381-95. 1994.

Brown, L.R. and Kane, H. "Reassessing the Earth's Population Carrying Capacity," *Full House*. New York: W.W. Norton & Company, 1994.

Cohen, M. *Hunger 1995. Transforming the Politics of World Hunger*. Washington DC: Bread for the World Institute, 1995.

Dreze, J. and Sen, A. *Hunger and Public Action*. Oxford: Clarendon Press, 1989.

Figure 2.5 **Food-aid trends: Predicting the food-aid balance**
Recent studies suggest that the availability of food aid will increase over the coming years at a much slower rate than the amount of food aid potentially needed. There will be less to go round. Agencies shipping and distributing food aid may have to become more discriminating in how they target this increasingly valuable resource.

	1996		2005	
	Low imports	High imports	Low imports	High imports
Global grain food aid	8.44	8.44	10.60	10.60
Chronic food-aid needs	33.20	29.70	42.20	34.10
Emergency food-aid needs	5.50	4.80	6.12	5.69
* The "1996" food aid figures are in fact 1994/95 year figures				

Source: ERS, US Department of Agriculture

Forster, P.W. "Tackling The Causes of Undernutrition in the Third World," *The World Food Problem*. Boulder, Colorado: Lynne Rienner, 1992.

Lappe, F.M. *Diet for a Small Planet*. New York: Ballantine, 1991.

Macrae, J. and Zwi, A.B., eds. *War and Hunger. Rethinking International Responses to Complex Emergencies*. London: Zed Books, 1994.

McCalla, A. *Agriculture and Food Needs to 2025: Why We Should Be Concerned*. Washington DC: Consultative Group on International Agricultural Research, Sir John Crawford Memorial Lecture, 1994.

Messer, E. and Uvin, P., eds. *The Hunger Report: 1995*. Langhorne, Pennsylvania: Gordon U. Breach, 1996.

Mitchell, D.O. and Ingco, M.D. *The World Food Outlook*. Washington DC: World Bank International Economics Department, 1993.

Newman, L. et al, eds. *Hunger in History*. New York: Basil Blackwell, 1994.

Pinstrup-Anderson, P. *World Food Security Trends and Future Food Security*. International Food Policy Research Institute Food Policy Report, Washington DC, 1994.

Sen, A. *Poverty and Famines: An Essayy on Entitlement and Deprivation*. Oxford: Clarendon Press, 1981.

Special issue on the Chiapas Mexico and the Zapatista rebellion, *Cultural Survival Quarterly*, Spring, 1994.

Uvin, P., ed. *The Hunger Report: 1993*. Langhorne, Pennsylvania: Gordon & Breach, 1994.

UNICEF. *The State of the World's Children: 1995*. New York, 1995.

Another component is the food needed to feed refugees and displaced people.

Commercial import capacity is a measure of the foreign exchange a country has available to finance food imports. The "high-commercial-imports" scenario assumes relatively higher growth in a country's total exports, in foreign exchange earnings, and in its capacity to import food commercially. The alternative "low-commercial-imports" scenario assumes slower growth in commercial food imports, based on lower export earnings which are associated with slower growth in global exports.

Clearly the implication for aid agencies is that food aid will become a more scarce and hence a more targeted resource. Programmes which use food aid to trade for cash are likely to become less common as food aid is increasingly seen as a nutritional resource only. Agencies may come under pressure to better target the food aid they distribute to ensure that it alleviates the most acute cases of malnutrition. Increasing the efficiency and effectiveness of the way food aid is used will be the hallmark of the next decade, particularly as agencies try to reduce the detrimental effects poorly targeted food aid can have on local food production and marketing systems.

Prescriptions for what should and can be done involve both short- and longer-term efforts. They also involve linking many different types of institutions and multiple development goals to initiatives to overcome hunger.

A 1995 study recorded progress in cutting hunger by half in certain Asian nations and significant efforts against micronutrient malnutrition in most of the world. Under UNICEF's leadership, major strides were being taken to expand maternal and child health care and child survival even in areas where economies were not growing.

For the longer term, the challenge of ending the "harder half" of hunger remains. Three "levers" to reduce world hunger were identified at the mid-decade: women's education, clean water, and better community organisation and infrastructure to facilitate progress. These move the mechanisms for food security away from a narrow vision of agriculture and nutrition towards links to other sectors, forces and actions.

Such linkages are also part of IFPRI's longer-term food security strategies. IFPRI, in developing its "2020 Vision for Food, Agriculture, and the Environment" focuses on generating investment for international agricultural research. But its discussions touch on many issues, from women to water, population to peace.

All these efforts to eliminate hunger and assure food security are part of a larger, global effort to involve communities, grassroots organisations and NGOs as true partners in development along with governments and intergovernmental agencies and actors.

Surveying these linkages and institutional advances, all attempts to recognise, frame and respond to hunger problems, the future of hunger still will depend on individual human choices. Will humans the world over choose family planning and so limit population growth and threats to sustainable food supply? Will individuals in the many societies of the world choose an environmental ethic that limits rapacious destruction? Will cultural food preferences lead individuals to reject richer but more wasteful diets, particularly where these damage sustainable access to food or good nutrition for all? And will affirmation that food is a human right lead to actions to assure minimum food security for all, now and for the future?

These are the important questions that will shape future hunger and food supply. They will determine whether we will all live in a global community that recognises that hunger is unnecessary and shameful for those who could prevent it, and practices as well as promises human rights and humanitarian values by meeting basic human needs for food and dignity. ∎

Roadside assistance: Food feeds the body of the refugee or the displaced person, but it can do much more and needs to do much more if agencies want to ensure good health not just calorie intake. Food can be a gesture of friendship and solidarity, it can help build for the future, it is a resource to be traded. Innovative uses of food aid, mixing food with other forms of aid, and efforts to better match resources to the needs of individuals and families, can conserve or enhance the capacities, security and capital of vulnerable people.

Feeding displaced people, Serbia, 1995. Abbas/Magnum

Food for lives or livelihoods?

Most refugees in poor countries have nutrition problems. Many are wasted, too thin, and a lot of these are hungry; some are not, because they are too sick to have an appetite. Many more suffer from micronutrient deficiencies, such as scurvy and pellagra, diseases which had nearly vanished by the 1970s, reappearing with the upsurge of refugees since then.

Total numbers of refugees and displaced people worldwide are estimated at more than 40 million. The largest group is in sub-Saharan Africa, where several million are at high nutritional risk. The other large group results from the Afghan conflict. The current estimates are that around 25 per cent of the refugees and displaced in the Afghan region are malnourished.

Almost all these people are displaced by war. The numbers, therefore, fluctuate with political changes, and as of February 1996 it seems that the size of the population affected in Africa had fallen from a peak of around 19 million in August 1994, to around 12 million in October 1995, and was now

Figure 3.1 **Nutritional risk: Refugees and IDPs nutritionally at risk in Africa**
In a continent where the normal population has a high risk of malnutrition, refugee and displaced persons suffer an even higher risk. Infant and child mortality is ten times higher within African refugee populations than in the rest of the continent.

Source: ACC/SCN, RNIS #14, February 1996

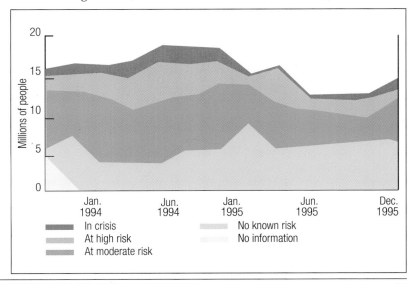

increasing again. Much of the reduction was due to resolution of the conflicts in Angola and Mozambique; numbers affected in Southern Sudan were also revised downwards at the end of 1994. This population, averaging around 15 million, is larger in size than all but six countries in sub-Saharan Africa, roughly the size of Ghana.

In this refugee "nation" of 15 million, infant and child mortality is at least ten times higher than elsewhere in Africa, accounting perhaps for a fifth of the total child deaths in Africa. Prevalences of wasting typically range from around 10 per cent to as high as 60 per cent in young children, and this quite closely correlates with mortality rates. Wasting is usually related to food availability. Results compiled over the last three years indicate that in half the cases observed the food availability was less than 1,500 kcals/head/day – less than three-quarters of a rather low estimate of requirement. Mortality in these groups went up to more than 20 times normal. Micronutrient deficiency diseases add to the burden of malnutrition.

It is especially tragic that a considerable part of the malnutrition, hunger, and mortality occurs in people within reach of help. Little may be possible for populations cut off by war or remoteness, while sick and severely malnourished new arrivals may make it hard to prevent raised mortality in camps in the first few days after their registration. But severe malnutrition and high mortality rates may continue for weeks, months or even years after refugees have arrived in accessible camps. There are situations where the influx is so enormous, and the conditions so hostile, that there is an initial catastrophe followed by improvement. A clear example of this is what happened in Goma. But if agencies or governments provide the right resources and organisation for displaced populations – even very large numbers in extremely poor conditions – the situation can rapidly improve.

Relief failures too often allow prolonged periods of malnutrition and death. This chapter focuses on how nutrition, hence health and survival, can be better and more rapidly protected and improved once populations of displaced people are accessible to external assistance, usually in refugee camps.

There is a widespread feeling that it is not only a tight supply of food aid that constrains more effective nutritional relief, but equally the efficient use

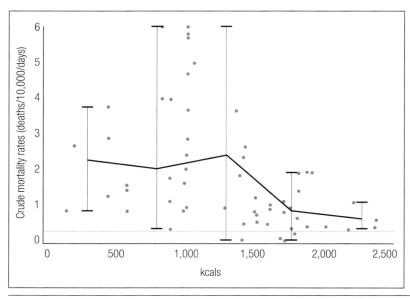

Figure 3.2 **Calories and mortality: The linkage between food intake and mortality**
In a recent survey of refugees, a clear association can be seen between increased food intake and decreased crude mortality figures. Malnourished people are more susceptible to disease and take longer to recover from illness. Ensuring that refugees and IDPs have an adequate diet also helps reduce the cost of health services that need to be provided.

Source: Database for RNIS reports

of available resources. While it is undoubtedly important to ensure a food availability of around 2,000 kcals/head/day, it is the quality of the food supply, the rational organisation of feeding programmes, the ability to deal with specific deficiencies, and the like which, on the ground, could make a big difference. There is a worrying lack of clarity concerning the underlying policies.

The fundamental problem causing hunger and malnutrition among refugee and displaced populations is destitution. Addressing nutrition while ignoring destitution is often ineffective and can cause continued high levels of malnutrition and mortality. Needs just as urgent as food – water, shelter, fuel, clothes – will force destitute people to sell or barter food to meet the cost of other essentials.

When agencies or donors calculate and provide food to meet only nutritional requirements, destitute people will inevitably become malnourished because they must use part of the food to secure other essentials. Giving additional food for trade, or by supplying non-food items or cash, will ease this problem. It is extraordinary that these problems persist, and that some donors resist offering more by implying that homeless people do not appreciate the charity of food because they use it to meet their family's needs for

Box 3.1 Common micronutrient deficiency problems

Vitamin A deficiency (VAD): Occurs in VAD-endemic areas, especially South Asia, Eastern and Southern Africa, causing blindness and increased mortality and morbidity risks among newborns and young children. VAD may be exacerbated by acute episodes of infectious diseases, especially measles, and is associated with increased case-fatality rates for those diseases. Most refugee rations provide less than the minimum RDA of 2500 IU per day.

Iron deficiency anaemia (Hb<10 g/dl in children aged 0 to 5 years and pregnant women; <11 g/dl in non-pregnant women; and <12.5 g/dl in men): Occurs worldwide, especially affecting women of child-bearing age, contributing to high maternal mortality rates.

It is common particularly where animal products (notably meat) are lacking in the diet. Iron deficiency may also be common in infants and children, and anaemia may be exacerbated by infection with malaria in endemic areas. In any event, this is probably the most highly prevalent micronutrient deficiency disease.

Scurvy (vitamin C deficiency): Epidemics of scurvy have mainly occurred in long-stay refugee camps in the Horn of Africa (Ethiopia, Somalia, Sudan and Kenya) where refugees have not had access to fresh fruit and vegetables. While scurvy might affect both children and adults, studies have shown that the risk is higher in women, especially pregnant women, than men, and increases with age.

The most telling risk factor is duration of stay in a refugee camp, reflecting the time on rations lacking in vitamin C. While the minimum RDA of vitamin C is 25 mg, between 10 to 15 mg daily will probably prevent clinical manifestations of scurvy.

Pellagra (niacin deficiency): This has occurred in refugee populations mainly in Southern Africa – in Malawi, Zimbabwe and Angola – where the cereal provided in food rations has been maize. Pellagra outbreaks have occurred when legumes, such as groundnuts, have not been provided in the ration.

While affecting both adults and children, women have been at far higher risk than men, and risk increases with age. Pellagra has been rare in infants and young children. At least 13 mg available niacin equivalents should be provided daily in food rations.

Beriberi (thiamine deficiency): This deficiency has occurred in refugee populations in Thailand, Nepal and Djibouti, resulting from cereal-based (especially rice) diets lacking in diversity. Thiamine deficiency in infants may cause high output cardiac failure and death.

Iodine deficiency disorders (IDD): Exists in most regions of the world, resulting from low iodine intake in the diet. Consequences include goiter, reduced mental function, cretinism, increased rates of stillbirths, abortions and infant deaths. Many refugees in Africa are located in countries where there is a high risk of IDD and no active control programmes. Iodised salt should always be used in rations to prevent IDD. ∎

Source: extract from M. Toole, 1994

shelter or clothing. Meeting both food and other needs is a fundamental policy issue whose resolution could significantly reduce malnutrition and mortality among refugees and displaced people.

There are other important issues offering far-reaching potential benefits for nutrition and survival. These include the overall composition of a food basket, its cultural acceptability, palatability and variety for nutrient needs. Clearer policy agreement is also essential for progress on types of feeding programmes; supplementing and fortifying diets to meet specific micronutrient requirements; and improving practices through better training and information.

Three ideas are put forward here. The first concerns monitoring to ensure that some acceptable standards of nutrition and health are met, and to trigger action if not. The second looks again at the question of meeting non-food as well as food needs, and the role of food aid in this. The third idea suggests that imaginative use of technology could not only lead to better "tailoring" of supplied foods to individual needs, economising on resources; but that this could begin to obviate the need for many of the complicated, often ill-considered, and time-consuming variety of special feeding programmes that, in reality, compensate expensively for other failures in the nutrition system.

Trigger levels

Standards will be crucial to the future of humanitarian relief. The *Code of Conduct* (Chapter 13) offers general principles of practice. Health and nutrition need similar ideas. Whatever the intention, the concept of acceptable minimum goals – on refugee mortality, for example – faces difficulties if taken to imply acceptance of anything less than normality. For example,

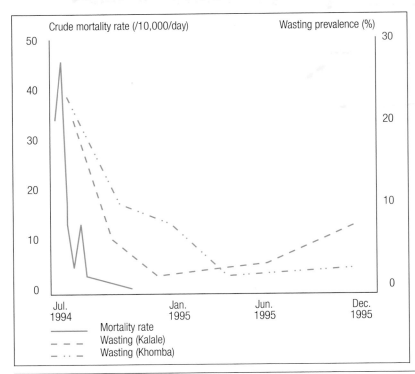

Figure 3.3 **Assistance and survival: Trends in wasting prevalence and crude mortality rates in camps for Rwandan refugees, Goma, Zaire**
Where displaced populations can be assisted, properly organised aid can have a rapid effect on mortality figures. Mortality rates amongst Rwandan refugees were cut rapidly over a period of a few months once aid started to flow and appropriate health services were provided. Likewise, malnutrition rates dropped dramatically over the latter half of 1994 as food aid distributions became more organised, but started to increase again with economic difficulties.

Source: Database for RNIS reports

setting an acceptable goal of mortality of 1/10,000/day, which is around three times normal, risks implying that a certain number of preventable deaths are acceptable, although this is not the intention.

A better concept may be "trigger levels", meaning levels of indicators requiring special action, over and above that already being taken, and possibly leading to heightened accountability. The actual action required would have to be decided on a case-by-case basis, but would almost always involve one or more of increased supplies of food, better health services, sanitation, water supply, fuel, shelter, etc. Because outcome indicators are the most commonly available – e.g. nutrition, mortality – these indicators might be particularly useful.

The Refugee Nutrition Information System (RNIS) started under the UN Sub-Committee on Nutrition (ACC/SCN), has shown that a surprising amount of data is regularly available, from agency reports, administrative sources, and specific surveys. Three of the most commonly available types information are: food supply, generally calculated in kcals/head/day; prevalences of wasting in young children; and mortality rates.

We are beginning to have reasonable experience of how these relate to each other. Calorie supplies below 1,500 kcals/head/day are often associated with mortality rising substantially. Above 2,000 kcals/head/day mortality is much lower. The prevalence of wasting in young children of 10 to 15 per cent is highly predictive of mortality rates above 1/10,000/day.

For mortality levels, the US Centers for Disease Control have suggested a convention that 1/10,000/day be regarded as very serious, and 2/10,000/day as a crisis out of control. Given the ethical issue of acceptability,

Box 3.2 Relative, absolute – what is an emergency?

"The meaning of emergency is a relative notion, very much depending on who looks at the process. For the people themselves, there is an emergency when they have to start selling the assets necessary for sustainable livelihoods, since their renewable resources are not any more sufficient to sustain a proper life. What they fear most is destitution. For the delegate of an international agency, it depends on how he feels about the question. For the politician, it depends on political interest. For the common western fellow, it is when people are starving to death. Furthermore, an emergency cannot be defined as an absolute set of conditions.

"The problem of an emergency probably lies in the fact that the people who are caught in the process depend on the will of others to help them, these others being in a position to decide up to what point they intend to help, in which way and with what kind of objectives. It is further complicated by the moral and ethical aspects which are not at all clear. Eventually, political considerations are in many circumstances a critical influencing factor on assistance.

"It seems clear that the more people are destitute, and if food aid is the only measure taken to help them, the more they will divert it since food, by the virtue of its exchange value, is likely to be their only economical resource.

"An important observation has been made regarding food consumption and poverty: the poorest households do not spend more than 80 per cent of their budget on food, even if they have to be malnourished. Indeed, if food is the only income that people have, as may be the case in camps, they will use part of it as a means of exchange to meet other needs essential to their survival.

"This was a striking fact in Ethiopia in 1985 and in Rwanda in 1992 and 1993, where people were completely destitute, and despite high rates of malnutrition and mortality, they were selling part of the food they received, at the high indignation of the humanitarian agencies. Needless to say that if the ration was already not sufficient to meet the energy requirements, malnutrition had to be an issue.

"Taking into account what has been said above about vulnerability, emergency and food in the context of famine, the obvious consequence is that one should advocate for an early and differentiated humanitarian assistance action." ∎

Source: extract from A. Mourey, ICRC, 1994

two trigger levels would be useful: crude mortality greater than the normal 0.3/10,000/day should require additional action; above 1/10,000/day should be regarded as a crisis demanding urgent intervention.

Trigger levels and related warnings are sometimes compared to indicator lights in mechanical systems. For example, green is adequate, an amber light a warning, and a red light a crisis. We probably need to add a further type – which we might call a "flashing red light" – to deal with continuing crises. There is a clear difference between a situation at the start, where all the indicators are red (e.g. mortality rate greater than 1/10,000/day, wasting greater than 15 per cent, kcal supply less than 1,500 kcals/head/day), which may be quickly brought under control; and a persistent emergency. This is the real opportunity for improved action.

A "flashing red light" should be turned on when (for example) there are two monthly reports consecutively above such trigger levels – greater than

---

Box 3.3 Nutritional indicators, cut-offs and definitions

Wasting is defined as less than -2SDs (standard deviations), or sometimes 80 per cent, weight for height (wt/ht) by the WHO/US National Center for Health Statistics standards, usually in children of 6 to 59 months. For guidance in interpretation, prevalences of around 5 to 10 per cent are usual in African populations in non-drought periods. More than 20 per cent prevalence of wasting is undoubtedly high, indicating a serious situation; more than 40 per cent is a severe crisis.

Severe wasting can be defined as below -3SDs (or about 70 per cent). Any significant prevalence of severe wasting is unusual and indicates heightened risk. Data from 1993 and 1994 show that the most efficient predictor of elevated mortality is a cut-off of 15 per cent wasting. Equivalent cut-offs to -2SDs and -3SDs of wt/ht for arm circumference are about 12 to 12.5 cm, and 11 to 11.5 cm, depending on age.

Oedema is the key clinical sign of kwashiorkor, a severe form of protein-energy malnutrition, carrying a very high mortality risk in young children. It should be diagnosed as *pitting* oedema, usually on the upper surface of the foot. Where oedema is noted, it usually means kwashiorkor. Any prevalence detected is cause for concern.

A crude mortality rate in a normal population in a developed or developing country is around 10/1,000/year which is equivalent to 0.27/10,000/day (or 8/10,000/month). Mortality rates can be given as "times normal", i.e. as multiple of 0.27/10,000/day. The Centers for Disease Control have proposed that above 1/10,000/day is a very serious situation and above 2/10,000/day is an emergency out of control. Under-five mortality rates (U5MR) are increasingly reported. The average U5MR for sub-Saharan Africa is 181/1,000 live births, equivalent to 1.2/10,000 children/day and for South Asia the U5MR is 0.8/10,000/day.

Food distributed is usually estimated as dietary energy made available, as an average figure in kcals/person/day. This calculation divides the total food energy distributed by population number irrespective of age/gender (kcals being derived from known composition of foods); note that this population estimate is often very uncertain. The adequacy of this average figure can be roughly assessed by comparison with the calculated average requirement for the population (although this ignores maldistribution), itself determined by four parameters: demographic composition, activity level to be supported, body weights of the population, and environmental temperature; an allowance for regaining body weight lost by prior malnutrition is sometimes included. Formulae and software given by James and Schofield (1990) allow calculation by these parameters, and results (Schofield and Mason, 1994) provide some guidance for interpreting adequacy of rations. For a healthy population with a demographic composition typical of Africa, under normal nutritional conditions, and environmental temperature of 20°C, the average requirement is estimated as 1,950 to 2,210 kcals/person/day for light activity (1.55 basal metabolic rate).

Indicators and cut-offs indicating serious problems are levels of wasting above 20 per cent, crude mortality rates in excess of 1/10,000/day (about four times normal – especially if still rising), and/or significant levels of micronutrient deficiency disease. Food rations significantly less than the average requirements as described above for a population wholly dependent on food aid would also indicate an emergency. ∎

---

15 per cent wasting, and/or greater than 1/10,000/day mortality, and/ or inadequate food supply.

Two points arise: are such data really available, and would the data trigger a warning too often? Data availability is not a major constraint and could – if there was demand – be improved relatively easily. Agencies send 10 to 20 new surveys each month on African refugee and displaced populations to the RNIS. Probably more are carried out and not reported. The most common measurement is wasting in pre-school children. Crude mortality rates are also quite common, usually from administrative sources rather than surveys.

The urgent action trigger level suggested would be roughly equivalent to the population categorised in the RNIS as high prevalence (in crisis) or high risk. These typically represent 10 to 20 per cent of the affected population, and this seems a reasonable estimate of how often an urgent intervention would be triggered. Perhaps in half the cases – say 10 per cent of the population – the situation persists, triggering a "flashing red light".

Food for non-food needs

The international donor community tries to meet what it defines as the humanitarian needs of war-affected individuals. Donors seem to have difficulty recognising that "humanitarian" or "basic" needs go well beyond food. Donors certainly do not see cash as humanitarian; it is believed that destitute people misuse cash for non-essential luxuries, such as alcohol, tobacco, or foods other than staples. This persists even in the developed world: the US government continues to distribute food stamps yet giving cash would bring large cost savings.

The suggestion that impoverished people are inefficient resource managers – blaming victims for their plight – is simply incorrect. The priorities of the poor may not be those of the wealthy, nor in line with donors' perceptions of their legitimate needs, but that does not mean they are wrong. Differences in perception, easily turned into moral judgements, cause many problems.

Agencies supply non-food goods to refugees and displaced people, from shelter materials to cooking pots, but distributions may be infrequent. Displacement may last ten years, and distributions of clothes or blankets are unlikely to match recurrent needs. Fuel for cooking and light is rarely provided. To pay for fuel, clothes, job seeking, family obligations, religious observances and even just the condiments or flavourings to make an unchanging diet palatable, trading food – the sole resource distributed regularly – may be the only option.

A food-use survey in Kanganiro camp in Burundi in late 1995 indicated that people ate two-thirds of the food ration and sold or traded one-third for non-food items or other foods, generating some 65 per cent of the refugees' total income.

Food provided to poor households usually improves nutrition because it is an additional resource; they either consume the food for direct nutritional benefit, or trade it for other foods. What is important, however, is that both ways lead to nutritional improvement. For war-affected populations dependent on external assistance, food aid also helps generate cash for essential non-food items. In such situations, the income transfer from the cash or barter value of the food is more important than nutritional content.

The World Food Programme does select food-aid commodities for their income transfer value in development projects where the function of the distributed food is to increase household income, as in food-for-work projects. It selects food commodities for emergency operations, however, only for their nutritional value.

Providing for the non-food needs of war-affected people with food is very expensive. Many individuals selling small quantities of cereals, grains or pulses at the same place and time may obtain very low prices. The cost effectiveness of transferring income with food aid – referred to as the alpha value – can be calculated by a simple equation: V/C, where V equals value to the recipient and C equals cost to the donor, including purchase, transport, administration and storage.

In Malawi, the food-aid ration for Mozambican refugees provided approximately 2,000 kcals/head/day, and the monetary value of the ration (using prices in government stores) was similar to or above the minimum wage earned by a Malawian household. As refugees received only food (and occasionally soap), they had to sell some food to buy any additional essential items.

Recipients did not consider valuable the commodities distributed to prevent an outbreak of pellagra, caused by a vitamin deficiency. For example, they did not distinguish groundnuts from beans, though the groundnuts (distributed for their nutrient content) were far more expensive. Vitamin B-fortified maize meal was sold by refugees at one-seventh of what it cost to get it there. Prices received by refugees were considerably less than the cost of the same commodity in government stores. Refugees needed firewood, soap, water, clothing, and additional foods, and were willing to sell at low prices.

If donors had provided the cash equivalent of 50 grams of the maize meal – a ration of 400 instead of 450 grams of maize meal – refugees would have received as much cash as if they had sold 350 grams of maize meal and the donors would have saved 11 per cent of their total costs. Refugees would have had more food to eat, more money or both, while donors feeding a million people for ten years would have saved millions of dollars.

There are many similar examples. Vegetable oil is a very expensive commodity for donors as well as commanding high prices in most countries. In Malawi it had the best alpha value of ration foods. Yet on the markets in camps in Ethiopia, Sierra Leone and Liberia, refugees sell vegetable oil for

| Commodity | Local price = V | | WFP cost* | Alpha value |
	Market	Store	C	V/C
Maize meal yellow**	56	—	397	0.14
Maize meal white	139	366	—	—
Mixed beans	200	635	657	0.30
Groundnuts**	200	—	1107	0.18
Edible oil	1,081	2,186	1,207	0.89
Sugar	—	385	607	0.63
Salt	—	500	357	1.40

* C.i.f. prices; includes resourcing, shipping, and internal transport, storage and handling
** Donated maizemeal fortified with B-vitamins as well as groundnuts were included to prevent pellagra
Note: V refers to market value except for sugar and salt

Figure 3.4 **Selling food aid: Food trading on the local market**
Refugees often have to sell part of their ration to buy much needed goods: soap, clothes, spices and the like. But often the price they get on the local market for their food is way below the value of the food. Such trading wastes both the nutritional value of the food and its cash value. Far better to either provide the extra inputs that the refugees need, or to provide extra food which has a high local market value.

Source: Database for RNIS reports

half its cost, and the same is true of fortified blended food, another expensive commodity distributed for its nutritional content.

Practical as well as humanitarian realities mean that additional provision of non-food resources including cash would be better for both donor and recipient. When this is impossible, trading of food, especially food with a high local value, is the only option. If cash is not provided, food-aid planners

---

Box 3.4 Displacement, technology, food – what next?

Camp diary, 4 June, 2021: When we arrived over the mountains, the first thing we did – after getting a hot meal, milk for the baby and my husband's arm bandaged – was sort out the family Relief Card.

A photograph was taken for the card, our details – height, weight, medical background, that sort of thing – were loaded on the card's microchip, and a record made of our hand prints for security controls.

The Relief Card is the key to food, water, fuel, shelter and cash. In the early 21st century it doesn't make being one of 100 million displaced people much better than it must have been to be a refugee in the late 20th century, but at least technology gives us some control over our lives.

The first thing we got from the card – after our details had been bounced to the central tracing database – was a message that our oldest boy, who got lost in the chaos of leaving, was safe in the camp down the road. So we began with a celebration.

The Relief Card is like a standard credit-identity card, but instead of being linked to a bank, it's linked to the camp database, giving me a weekly level of credit for food and water, fuel and shelter, based on my family's needs, and access to health care, education and other services.

I can use it to "buy" items from any of the camp stores or to "rent" services – a house, electricity, piped water – at market rates from camp sector managers. The camp stores are run by a range of aid agencies, local and international companies and a camp members' cooperative, so I have a choice of what, where and when to buy.

Private firms do most of the delivery, competition keeps prices down and quality up, and it's easy to swipe the card through the machine and put your hand on the security plate. Queues? Not like the food distributions in old camps, I'm told, at least not when you go through "ten items or less".

The Relief Card has 10 per cent of its value as cash, so I can make purchases outside the camp stores for travel, say, or a phone call. I could use the money to start a small business. I can even pay into the account from my earnings. Once I have a payment track record, however small, further credit is available at the prevailing world rates. Best of all, I don't have to sell food to raise money.

Security on the cards is tight and they are hard to fiddle. The card has the family hologram on it, and each purchase needs an adult there to be identified by hand print recognition. Most people have their regular stores, so the wrong card or the wrong person is quickly spotted.

At the health centres, the card's medical records can be checked and updated. I was told that one benefit is eliminating the massive cost and complexity of supplementary feeding centres. If a kid is malnourished, the family ration goes up for a while with monitoring at the health centre.

Of course, the kids don't much like the way the sugar ration is cut if they need too many fillings at the camp dental centres, but it certainly makes them clean their teeth. And they soon get good at making credits through the recycling schemes, and once old enough, by working on the camp environmental projects.

The camp has lots of other technical stuff: satellite pictures help map sites, check camp density and maintain security; the tent-houses offer good insulation with new materials but are light enough to be carried on your back if the fighting gets close; and all the education and training on interactive video and audio comes via low-orbit satellite communications from the Internet.

Being displaced shouldn't grind you into poverty; indeed, although we lost a lot when we left, I think by careful saving, hard work and taking full advantage of everything on offer we should come out with more than when we arrived.

As everyone knows, this isn't charity, it's a human security investment with a calculated payback in lowered future costs from greater stability and less conflict, poverty and disease. Hungry, angry and poor people spell trouble, so while the negotiations and arguments go on about the conflict, we are encouraged to stay fit and healthy, get educated and be economically active.

After all, who in the world needs people without purchasing power? That's no way to run a global market economy. ∎

---

and donors must accept that food will be traded and not take evidence of trading of food as a reason to reduce the amount of food provided. Fuzzy thinking and muddled ethics are standing in the way of more effective use of limited financial resources

Three factors are required to ensure the adequate nutritional status of a population, especially that of children. Enough food of good quality must be available; young children must have adequate care; and infectious diseases must be controlled through sanitation, hygiene, immunisation and other preventive health steps, and medical treatment. Insufficient attention to any one factor increases child malnutrition. In poor populations, this means more children die.

Nutritionally efficient food aid

Sufficient food should be provided to meet the nutritional needs of the entire population. Food can never be distributed to a large population in a totally equitable manner: some people get more and some less. It is usually considered impractical to meet the differing needs of individual families. Agencies usually plan food distribution programmes by assuming that all families and individuals have equal food needs.

Difficulties in obtaining accurate information about the number of people needing food, problems of food availability, delays and losses can cut the total amount of food entering a camp to 75 per cent or less of the planned food requirement. As a result of this some individuals will then actually received only half of the food they need. Such individuals will often include the most vulnerable – young children and pregnant women. Female-headed households will also often be at a particular disadvantage when the overall supply of food is insufficient. Moreover the effort that families make trying to find ways of acquiring additional food to alleviate hunger reduces their ability to care adequately for their most vulnerable family members, further increasing the numbers of malnourished.

In this situation those concerned with child well-being may urge an increase in the general ration. Donors or agency headquarters may find it hard to accept the need for an increase in the total quantity of food, particularly if they also receive reports that considerable quantities of food are being traded. Even if donors accept the need for more food, extra supplies may take months to arrive. Faced with shortfalls or delays, those with particular responsibility for children will typically set up a parallel food distribution system, relying more on "supplementary feeding programmes".

In these supplementary feeding programmes children are weighed, mothers given nutrition education and a ration of special food, often a blended fortified cereal. Sometimes this approach may be the only one feasible and operate effectively; but often such programmes do not work well. Mothers wait for hours in line, individual children get inadequate attention, and counselling is inappropriate. The extra ration may also be too small to improve significantly the nutritional status of the target child, once shared with the rest of the family.

There are two potential solutions to this problem. Agencies could adjust the size and composition of the standard family food ration to meet the needs of the family with the greatest food requirement. That would require substantial increases in total requirements. Food-basket composition could be adjusted to meet the particular needs of each family; individual tailoring of rations may be worth exploring.

While the practicalities of food-aid distribution have changed little in decades, the technologies of the supermarket, security industry and satellite navigation, linked by the now omnipresent laptops of aid workers, offer new options. These could ease registration, nutrition targeting and monitoring,

Sources, references, further information

Desenclos, et al. *Epidemiological patterns of scurvy among Ethiopian refugees.* WHO Bulletin, Vol. 67, pp.309-316, 1989.

James, W.P.T. and Schofield, C. *Human Energy Requirements.* Oxford: Oxford University Press/FAO, 1990.

Moren, et al. *Outbreak of pellagra among Mozambican refugees – Malawi, 1990.* MMWR, Vol. 40, pp. 209-213, 1991.

ACC/SCN. *Update on the Nutrition Situation, 1994.* Geneva, 1994.

ACC/SCN, RNIS. *Refugee Nutrition Information System Bulletins.* No. 14, February 1996, Geneva.

UNICEF. *State of the World's Children.* New York, 1994.

and food delivery to individual families based on their specific needs. Planning, testing and preparation would allow emergency use of this technology.

Refugee situations present challenges in the design of a reliable rationing system. A major problem is the sale or bartering of ration cards, while families may borrow extra children to increase their entitlements at registration or re-registration. Improved monitoring and supervision systems are required. Technology is now available which can allow an individual to be uniquely identified by taking accurate hand measurements, and such methods deserve to be reviewed to see if they could help everyone to get enough food.

The same level of investment in information management technology that the average US supermarket makes could greatly facilitate food management in any large refugee camp. Photographs, bar codes, swipe cards and chip cards all offer new ways to streamline food-aid delivery, cut waste, improve targeting and reduce the risk of malnutrition.

Computerised records linked to security systems could be tried, while other technology, such as the satellite-based global positioning system might greatly improve camp mapping and location of family units, and facilitate survey work and monitoring.

Future options

The three ideas put forward here require only small additional resources, compared with the bulk supplies of food and other assistance provided. Yet they have a chance of significantly improving nutrition among the most vulnerable people in the world, saving lives and guarding health.

Information plays a key role. The idea of triggering action using quicker and more specific interpretation of existing data could focus limited resources on those most in need. It would improve accountability, and lead towards better codes of practice. Information technology could provide proper individual identification, the key to better distributing available supplies, tailored to special needs of the vulnerable.

Recognition that there are needs just as basic as food would lead to a more diversified – more humane – approach to humanitarian assistance.

As needs grow and resources shrink, donors and agencies must bring together new technology and new thinking in nutrition and food aid to bridge that humanitarian gap for many millions of vulnerable people. ■

Capacity checklist: Disaster survivors are much more than numbers. The vulnerable also have knowledge – especially of their needs, circumstances and survival strategies – and capacities. Even in the urgency of immediate relief work, there are ways of building on local capacities. As populations on the move grow, offering them portable resources and skills will be crucial. Particularly in relief programmes that last for months or years, a developmental approach is essential if opportunities for the future are to be kept open.

Salvadorean refugees, Honduras, 1987. Gilles Peress/Magnum

Meeting more than basic needs

Recent years have seen a massive rise in the amount of relief work undertaken by international humanitarian agencies separate from the governmental response structure, from United Nations (UN) agencies to the International Red Cross and Red Crescent Movement and non-governmental agencies (NGOs). In 1994, 3.5 billion US dollars were spent on humanitarian assistance for many millions of people and millions of tonnes of food aid were disbursed. An increasing proportion of this aid is being spent in huge relief operations to assist refugees, internally displaced people and those fleeing war.

Three features of such programmes make them particularly difficult to deal with in a humane way. First, they often start up quickly and in ways that create immediate and urgent needs. People on the move, regardless of whether they have crossed an international border, are cut off from their normal means of survival. They cannot farm, they cannot trade, they no longer receive wages or state benefits, they do not have access to a health system.

Second, the numbers of people involved are enormous and concentrated. It has become common to think in terms of assisting populations of hundreds of thousands at a time, often crowded in camps a few kilometres across. Such concentrations of vulnerable people breed their own problems. Communicable diseases become more prevalent. People fear intimidation and crime.

Finally, the cutting of links to the old, pre-crisis way of life is much more severe than in smaller disasters that leave community foundations on which to build. Assistance, for many who flee crises, is a holding operation, a resting place to recoup while waiting for "something to happen" so people can go home, move on, or start to settle where they are.

These massive relief operations represent a growing proportion of the world's humanitarian effort. They are also the most extreme, the most visible and the most media-hungry of events. They are the times when the humanitarian community feels under public pressure to get it right and demonstrate that it knows what it is doing. At the same time, there is the continuing need for relief assistance for lower-profile disaster events: floods, earthquakes, drought-induced famines, technological accidents. These may not affect so many people at a time or be so public, but responding to them professionally is just as critical.

Relief work is about the "bottom line" of ensuring basic minimal necessities to keep people alive. Survival is the minimum that relief must secure.

Basic relief attempts to secure survival by ensuring that people have access to four things: sufficient drinking water and sanitation; sufficient food; basic medical care; and shelter from weather. A fifth basic necessity is added in many of today's disasters: protection from violence and harassment. Assistance and protection underpin all relief efforts, but are they sufficient, should relief do more?

Traditionally the aid community divided itself into two camps: those concerned with development and those concerned with relief. Both camps had at least a tacit belief in a development model that saw developmental inputs leading to an ever-increasing standard of living, with disasters as temporary blips on the ascending development curve. Disasters were temporary, albeit tragic, phenomena that did not really affect the prospects for long-term development. Under such a model, which still shapes the thinking of many humanitarian agencies, development carries on around the disaster, while relief patches things up so that disaster victims get back up on the development curve.

This model works well for many disasters triggered by cyclical natural events, such as hurricanes, floods or droughts. It envisages a response cycle of preparedness, then disaster, then relief, then rehabilitation. Gradually,

Box 4.1 Uganda – going beyond refugee survival

At various times since independence, Ugandans and Sudanese have swapped the roles of refugee and host. Many Ugandans were refugees for years in Sudan and played an active role in relief work. Now home, their experiences condition their attitudes towards refugees and relief operations.

In northern Uganda, after a fresh influx of Sudanese refugees fleeing the civil war, it was decided to move up to 60,000 refugees into the sparsely-populated Aringa country where farming land was said to be available. Ikafe, the area in Aringa offered to the Sudanese, had a population of about 4,000. Apart from tsetse flies and scorpions, the real problem was lack of water, the reason for Ikafe's low population.

In December 1994, Oxfam UK and Ireland – hoping to break away from the "relief equals minimal assistance, minimal standards" mentality – commissioned a report on relief work at the Ikafe settlement. The report made five key points:

Integrated approach: Many aid officials believe that keeping refugees at the survival level will discourage dependency and hasten the end of the relief operation by encouraging them to repatriate. However, this view is mistaken. An integrated approach is necessary, where relief beneficiaries, from both host and refugee populations, are defined according to needs, not status. As part of an effort to win the support of local and refugee communities, relief agencies should consult their formal and informal leaders, as well as government officials.

Capacity building: Rather than developing parallel services, agencies should build up the capacity of local structures. Existing Ugandan legislation should be used to set standards for refugees and hosts.

Aiding livelihoods: Aid must support the refugees' own efforts to reconstruct their lives and communities. The Sudanese refugees, who are semi-nomadic, cattle-owning people, need mobility to find their relatives and create mutually supportive groups, rather than stay in the groups created by camp life.

Food distribution mechanisms, which aim at providing per capita assistance, and camp arrangements actively militate against this.

Economic action: The present assistance level keeps people alive but cannot boost the refugee economy through employment. Flourishing schools and religious groups would help refugees achieve a better standard of living, but this requires more than a basic survival package.

Political understanding: The project, like many relief efforts, has a tendency to ignore the complicated local political environment, leading to unnecessary conflict and confrontation. Resources can be used more effectively, confrontation avoided and trust built up if agencies seek to understand this environment and deal with it transparently.

Finally, the report urges more action through local government, local NGOs and local refugee-based groups rather than involving more foreign NGOs in the project. ∎

rehabilitation activities merge into normal development activities; crisis over, life goes on. There is a continuum, with relief quickly replaced by rehabilitation, which is gradually replaced by development activities.

This model also shapes aid funding structures. Development funding is usually only available after painstaking needs assessment, problem identification, feasibility studies and the like, requiring detailed budgeting and implementation paralleled by tight monitoring and evaluation. Donors and agencies saw relief as a simpler process in which rapidity of action was paramount, needs assessment less important – because needs were obvious – and monitoring a luxury.

While such arrangements may not be ideal, they represent a sensible compromise where there is an assurance that relief will quickly pass into development. But what if the model is wrong? What if there is no immediate development process to return to, or relief becomes a long-term activity? Should this change conceptions about relief, and are there standards other than "meeting basic necessities" to judge the quality of relief work?

The linear model of development is under attack on two fronts. First, many development workers are questioning whether it was right. While development inputs have led to economic success, most notably in South-East Asia, elsewhere, especially in Africa and most tragically in Rwanda, development inputs seem to have had little effect on the quality of life of most people.

Second, relief workers are starting to question the model. They see that reality for many disaster victims holds no prospect of a fast return to normalcy and that receiving relief can rapidly become a way of life, particularly for refugees and displaced people in camps.

The reality of humanitarian assistance today is that many people are

Box 4.2 Bosnia – rebuilding with milk and potatoes

Like any sophisticated city, Tuzla in Bosnia depends on trade, employment and links with the countryside to survive and prosper. War shatters all of these.

In 1994, Norwegian Church Aid (NCA) took on the task of providing relief to the people of Tuzla. The approach challenged the notion that relief is a one-way process – from aid agencies to beneficiaries – and sought to save livelihoods, not just lives, by building on local capacities and supporting local institutions.

Tuzla is a market town in the heart of good agricultural land. In normal times, vegetables were always on sale and locally processed milk was available through schools and health clinics as well as normal commercial channels. NCA sought to provide assistance by helping the city restore milk and vegetable production.

Prior to the programme, distribution of too much free "relief-food" milk closed down local milk production. Dried skimmed milk is a surplus product of both the European Union and the United States. If clean water is available, it is a valuable relief commodity to meet nutritional needs at the right time and place. In Tuzla it destroyed the market for local produce, cut production, increased unemployment and reduced the chance of reviving the milk industry later.

NCA stopped distribution of imported milk powder and supplied spare parts to the local dairy. Through the dairy, it arranged with milk producers to buy 2,000 litres of milk every day for distribution to refugee and host children in the town. This brought many benefits in addition to meeting the nutritional need. The town's well-functioning dairy is ready to carry on once conflict ceases. Local dairy herds have not been slaughtered for cash. Relief needs are met and there is milk in local markets available for all.

To encourage vegetable production, NCA supplies fertilisers, pesticides and seeds to people in Tuzla and neighbouring Gradacac and Bosnaplod. In return, NCA has the right to purchase one-third of the harvest.

So far, as well as boosting the general availability of food, it has bought 3,000 tonnes of vegetables, fruit and cereals for distribution to newly-displaced people. ∎

dependent on relief, or at least have little chance of returning to normality, for extended periods of time. In Malawi, refugees spent years, some of them up to a decade, in relief camps before being able to go home to Mozambique in the early 1990s. Around Rwanda, three-quarters of a million people are still surviving in refugee camps two years after they fled the country, and within Rwanda tens of thousands of people remain dependent upon humanitarian aid for survival.

Long-term relief

The same story is repeated in the Caucasus, Liberia and Afghanistan. Equally, the drought-relief projects in East Africa and the flood-response programmes in China, Viet Nam and Bangladesh all assist disaster victims for whom relief has become a part of survival. Those survivors cannot wait for development to replace relief. Relief work itself has to become more developmental in its thinking, go beyond the five basic necessities and start to examine not just what it delivers but how it carries out the delivery.

How should we recognise a relief programme carried out in a developmental way? What is different about a "developmental" relief programme? In 1995, 85 humanitarian practitioners from some 15 countries, representing relief organisations, development agencies, government donors and UN bodies came together in Copenhagen to address these issues. Drawing on field experience of relief work in Africa, Asia and Europe, they sought common features that exemplify the developmental approach to relief.

They identified three key areas where developmental relief goes beyond basic relief. First, it seeks to communicate with beneficiaries. Second, developmental relief looks to sustain livelihoods, not just lives. Third, it aims to build on local realities. Each of these three categories includes three specific features, giving nine areas to improve the quality of relief.

Participation: Developmental-relief programmes deliberately involve disaster survivors in the decision-making process, aiming to empower them to retake charge of their lives. Even in particularly difficult situations, such as relief to large-scale displaced populations, diverse community leaders can help assess the situation and identify useful resources.

Accountability: In relief projects, agencies traditionally see themselves as being accountable upwards, towards their headquarters and donors, but they should also practice accountability towards disaster survivors. At a minimum, agencies should openly share information on the planning, execution and expected duration of the relief programme with beneficiaries.

Decentralised control: Programmes for developmental relief should take management decisions as close to the beneficiary population as possible.

Demonstrating concern for sustaining livelihoods: Developmental-relief programmes are concerned with what comes after a relief operation, and how it is carried out. They provide assistance that complements, rather than competes with, the normal means of livelihood of the disaster survivors.

Strategies based on the reality of the disaster: Relief programmes address many different types of disasters: those triggered by natural events, those which develop slowly over vast areas of a country, those caused by war and economic collapse. Strategies for developmental relief should be adapted to suit the environment of the disaster rather than relying solely on pre-packaged delivery derived from a model of only one type of disaster.

Identifying the needs and capacities of diverse disaster survivors: It is essential to recognise that survivor populations contain many groups with different capacities, vulnerabilities and needs. Programmes for developmental relief should seek to address these diverse groups and their capacities as well as their different needs.

Building on capacities: The need to access vulnerabilities is recognised as being important, but relief programmes that seek out and work with capacities, skills, resources and organisational structures of disaster survivors will be more effective than those that assume survivors are passive and helpless.

Building on local institutions: Imposed relief programmes can undermine local structures, which are often used, but not strengthened, and then abandoned after the relief operation. Developmental-relief programmes aim to work with local institutions and build their capacities to carry on humanitarian work after the need for relief has passed.

Setting sustainable standards for services: Relief programmes often set in motion the development of service and welfare systems – health, education, water provision and the like. These should be of a standard and provided in a manner that has a realistic chance of being sustained after the relief operation ends.

Between relief operations, aid workers and agencies agree that they should find ways of doing relief in a more developmental way, but when the crisis hits and fast decisions are made on minimal information, managers avoid the risk of using other than tried and tested responses.

The international aid system contains many actors, all of whom need to participate in changing the system's end product: relief delivery. Changes are necessary in the way aid agencies work, in the way donors provide funding and in the way the academic research community views relief work. Each can make changes with minimal investment. Below are suggestions for possible improvements in the quality of service each group can offer to disaster victims.

High standards

Programming standards: Agencies must set themselves high and defensible humanitarian standards if they are to practice developmental relief. As a starting point, agencies are encouraged to subscribe to the standards laid down in the *Code of Conduct for the International Red Cross and Red Crescent Movement and NGOs in Disaster Relief* (see Chapter 13).

Specialised competence and coordination: Large relief programmes attract many external relief agencies. To improve the effectiveness and quality of their activities, agencies should examine critically their own strengths and seek to develop greater competence in limited fields, if necessary, rather than a breadth of mediocrity in service delivery. Relief organisations should be willing to coordinate when it adds to the greater good of a project, and should recognise the need to balance their right to independence of action against the humanitarian value gained through coordination.

Programme reviews: Many relief programmes go on year after year in the same way. By reviewing projects annually, agencies can identify changes that progressively make more use of local leadership, skills and capacities.

Sharing experience with donors and the media: Agencies need to improve the sharing of field experience – of success or failure – with donors and engage in a dialogue about policy change. Equally they should work more effectively with the media to build understanding of the issues and to break down stereotypes such as those of "helpless disaster victims".

Linking relief and development programming: The present organisational structure and funding mechanisms of many donor institutions reflect the view of relief and development as two divorced activities. Donors should seek ways of promoting dialogue between their relief and development divisions, and of allowing a degree of development funding into relief programmes.

Accountability in quality: Measuring the quality of developmental-relief programmes requires a different set of parameters and associated skills from evaluating simple relief delivery. New ways of evaluating and reporting relief programmes which reflect a developmental approach need to be explored.

Support for local structures in relief and disaster preparedness: Working through, enhancing and supporting local structures is central to the developmental approach to relief. Donor institutions should recognise and support the legitimacy of funding and strengthening of local structures as part of disaster-preparedness and -relief programmes.

Developing practical methods of capacity and vulnerability analysis for disasters: Developmental relief places greater emphasis on understanding local capacities and vulnerabilities than needs-driven assistance delivery, yet few methodologies exist to help assess these features. Research bodies should develop methods of capacity and vulnerability analysis appropriate for relief situations.

Developing methods to evaluate quality: Measuring and evaluating the quality of developmental-relief programmes requires a different set of parameters and associated skills than evaluating simple relief delivery. In collaboration with agencies and donors, the research community needs to establish methodologies for appropriate evaluation techniques.

Developing accountability systems: Present relief accountability systems stress financial reporting supported by process-descriptive narrative. A more holistic reporting system needs to be devised to provide information on features of relief programmes. Such a system should also offer quantitative delivery information, such as capacity building, participation and accountability to the disaster survivors.

According to one seasoned relief worker, who has experienced the shortcomings of narrowly-defined relief in both Afghanistan and the Balkans: "Often external relief organisations and funding bodies set up parallel structures to already existing local ones. Once established, these new structures are hard to change.

Box 4.3 Information helps ensure rights for refugees

In August 1995, a military offensive in the Krajina region of Croatia sent 175,000 Krajina Serbs on a trek through west and north Bosnia into the Federal Republic of Yugoslavia. Like all refugees, they were cut off from normal means of survival. Like many refugees, they had witnessed the brutality of war. In common with disaster victims the world over, they needed immediate assistance to start rebuilding their lives.

The Yugoslav Red Cross, with support from the International Federation, provided help – from food to blankets, water to medical assistance – at border crossings, in transit centres and as people moved into collective and private accommodation. Over 8,500 Yugoslav Red Cross volunteers scoured the roads for individual refugees who had fled.

Most refugees arrived frightened and tired; many were angry. In such a situation, suspicion can quickly take hold. People need to know what is going on. No one is an expert at being a refugee and the people of the Krajina had become refugees overnight.

With the help of the Yugoslav Red Cross and the Federation, the refugee publication, *Odgovor*, compiled and published 41,000 copies of a booklet entitled "Guide for Refugees". Yugoslav Red Cross networks distributed them throughout the country.

The booklet's contents included a guide to the basics of humanitarian law, profiles of active humanitarian organisations, contact addresses and useful, simple health-care messages.

Information campaigns can change the nature of the relationship between the assisted and the provider. Those being assisted are no longer seen as helpless victims. If disaster survivors know what to expect from assisting agencies, there will be pressure on agencies to live up to those standards.

Public pressure ceases to be solely an issue for an agency's home base, but becomes a live issue in relief programmes, allowing much more equitable relationships to be built between the beneficiary and the relief agency. ∎

Sources, references, further information

Anderson, M.B. and Woodrow, P.J. *Rising from the Ashes: Development strategies in times of disaster.* UNESCO, Paris, 1989.

Maxwell, S. and Buchanan-Smith, M., eds. "Linking Relief to Development", *Institute of Development Studies Bulletin,* Vol. 25, No. 4, 1994.

"Where local organisations are used, external agencies rarely take the time to study their structures, motivation, management and governance. Local organisations are often asked to manage resources on a scale which is orders of magnitude greater than that which they are used to. This can lead to real problems where there is no parallel management assistance. Local organisations often find themselves spotlighted in the national and international media. If handled incorrectly, this can damage the standing of the local organisation with the government and community.

"Changing the way we do relief isn't a matter of academic interest, it's a matter of survival for many of the people we assist. They have a right to expect certain standards from us and we have an obligation to use donated funds given to us in trust to gain the maximum impact."

Doing relief in a more developmental way is a major first step in this direction. ∎

Box 4.4 Kenya – making participation work in camps

The first boat of refugees escaping from fighting in Somalia arrived at the Kenyan port of Mombasa on 16 January 1991.

Camps for tens of thousands of people opened along the coast north of Mombasa. In mid-1994, the Kenyan government decided to close the camps and to repatriate all refugees or move them closer to the Somali border.

The Kenyan refugee camps started as a typical relief response to an urgent problem. Through a social outreach programme employing suitable refugees, the Kenya Red Cross introduced a developmental edge to its relief work with refugees. One aspect of this work was an attempt to involve refugee groups in the design and implementation of relief and welfare work.

Refugee groups were asked to suggest projects by setting out how the idea would address an identified need, with its specific objectives, the resources needed, budget and evaluation process. Meetings of group leaders and social service staff would review projects.

In one camp, the leader of the women's committee was very well educated and had held a top post in the former Somali government. She recommended five projects, in this order of priority: a gymnasium; a beauty parlour; a wholesale business; a bakery; and child care.

A discussion was held with the women's committee leader and a group of women she selected to debate the proposals. They concluded:
● Only an elite in the camp was familiar with the idea of a gymnasium. The idea would be unknown to most camp women. Women would be unlikely to use gymnastic skills in the future;
● A beauty parlour would not meet the needs of very poor women;
● The bakery and wholesale business ideas needed to the explored further. They seemed designed to provide a large income for few women and would probably favour committee members.

The meeting agreed to make child care the first self-help project, with these benefits:
1. Training 15 women in child-care skills, since war had disrupted normal family care.
2. Preparing 300 children for future education.
3. Improving the children's nutritional status through feeding with high protein biscuits and milk.
4. Giving mothers free time to meet other needs.
5. Offering employment for trained women.

By involving refugees in this project's design and implementation, the Kenyan Red Cross was able to give a strong developmental edge to relief work. The project was a success and was later implemented in other camps. ∎

Funding application: The humanitarian "market" is changing fast, forcing change on its "stakeholders", from donors to the military, agencies to beneficiaries. Patterns of need and patterns of donor response have shifted greatly in the last few years and show every sign of continuing to change as a new century of crisis opens. Like them or hate them, humanitarian agencies need to understand these changes and anticipate them if they are to continue to provide high quality services to disaster survivors.

Bread queue, Grozny, Chechnya, 1995. Donovan Wylie/Magnum

The state of the humanitarian system

Profound changes have been taking place in the international relief system in recent years. In crises as diverse as Kurdistan, Rwanda, Liberia, Somalia and Yugoslavia, humanitarian assistance is no longer funded, carried out or judged as it was half a decade ago.

This chapter, which will be a regular feature of the *World Disasters Report*, examines policy trends in the international relief system, focusing this year on responses to conflict-related crises. While the financial costs of responding to crises continue to rise, there is a lack of clear objectives and values to guide and monitor the system. Defining those objectives and values will depend, in part, on confronting major uncertainties within the wider domains of foreign policy and aid policy.

The international aid system developed during the Cold War. The super-powers and their allies rewarded client states with financial assistance, and withheld support from opposition states. The decades following World War II were marked by decolonisation and state-building which, from a donor perspective, took place within a framework of clear ideological divisions.

Within this framework, questions about aid's economic and social impact led to competing schools of thought on the objectives and strategies of official development assistance. Despite an apparent competition of theoretical and practical approaches, considerable consensus remained on two fundamental tenets guiding official development assistance to developing countries.

First, a linear model of modernising development which proposed a progressive process leading to economic growth under the umbrella of a liberal, democratic political system. Disasters, including conflict-related emergencies, were perceived as temporary interruptions following which "normal" development could resume.

Second, despite increasing emphasis on the growth of the private sector, official development assistance remained dependent upon the presence and consent of strong state structures in recipient countries. For example, to receive loans and grants from the World Bank and the International Monetary Fund, an internationally recognised government must be in place to incur debt and define policy objectives.

Since the late 1980s, these tenets of aid policy have been shaken. In 1995, the United Nations (UN) identified 28 complex emergencies, affecting some 60 million people. Most of these emergencies were conflict-related. The persistent nature of conflict and its rising humanitarian and financial costs imply that disasters can no longer be seen as transitory phenomena.

The boundaries of the international relief system have changed markedly in recent years. Historically, they were demarcated in three key ways. First, separation of relief and development, marked by the aid agencies' specialist mandates and specific bureaucratic and financial procedures. Second, separation of aid agencies from political and military bodies, and a convention assuming aid's neutrality, particularly emergency aid. Finally, sovereignty's geographical and political borders: humanitarian action was limited by the willingness of governments to allow relief operations.

Such rules appeared to guarantee the neutrality of emergency aid – its purity of purpose, from the donor perspective, signalled by its lack of conditionality – and the mechanisms for accountability, making recipient governments responsible for defining their citizens' needs and allocating resources accordingly.

Changes had begun before the Cold War ended, as weaknesses in these approaches were revealed. After the superpower confrontation ended, concerns to protect neutrality and sovereignty were subjugated to promote new

---

Box 5.1 Media and resources – making the link

The history of humanitarian aid includes many instances where the media, particularly television, appears to have played a critical role in galvanising the international community into providing substantial resources. Ethiopia in 1984 and Northern Iraq in 1991 are two much-quoted examples.

The influx of 850,000 Rwandans into Goma in Eastern Zaire provides a recent and powerful case. Real-time satellite broadcasts began within one day of the start of the flood of refugees and a powerful story, strong images and a palpable sense of urgency were reinforced when a cholera outbreak began a few days later. A "media frenzy" developed drawing in scores of journalists and, from mid-July to mid-August, the story was given significant space in news broadcasts and print media.

The international community's response to the influx was exceptional, involving a large airlift, the arrival of up to 100 NGOs and deployment of military units. Before the influx, aid agencies found it hard to find funds to help people waiting across the border in north-west Rwanda or preparedness work in Goma itself. By late July everything from bulldozers to bottled water was being flown in.

Despite the apparent link in this case between the scale and intensity of media coverage and the response, it is difficult without rigorous academic analysis to determine precisely the influence of the media and how and why it can vary. Media coverage of human suffering has often not resulted in greater assistance levels, while in some instances, such as the 1991-1992 Southern African drought,

massive mobilisation of resources occurs despite limited TV coverage.

The story's emotional power, especially the television images, seems critically important in determining heightened media interest and how it is ranked against competing news stories. This in turn appears to strongly influence the extent to which the coverage affects resource allocation.

Ethiopia, Iraq and Goma involved not only large-scale and intense human suffering but also concentrations of affected people which dramatically increased the emotional strength of the initial coverage. While such concentrations usually reflect a major humanitarian emergency, this is not always the case – very high death rates have been experienced among populations which have not been displaced or gathered in camps or feeding stations.

This bias in the impact of media coverage towards situations which are more "mediatic" – concentrations of vulnerable people, when even immediate action cannot prevent loss of life – presents the humanitarian aid system with considerable challenges.

The media will play by their rules, but humanitarian agencies have to become more assertive in promoting concern for all disaster victims, regardless of the presence of the TV camera. Also, aid agencies have to lobby their donors to get the right balance between the free giving of funds and tight earmarking.

Agencies need the flexibility to spend cash where it is needed, not where it will be filmed. ■

international rights and responsibilities to manage conflict and its humanitarian crises.

Recent moves to reform the international aid system to fit today's turbulent realities have highlighted the issue of linking relief and development (see Chapter 4). The concept of the relief-development continuum had its origins in natural-trigger disasters, highlighting the links between prevention, preparedness, mitigation and recovery in reducing the impact of earthquakes and drought. Based on this approach, advances have been made in architecture, engineering, insurance and asset protection in many countries.

Evidence suggests, however, that the relative strength of the economy and of political and bureaucratic institutions in hazard-prone countries determines the effectiveness of such measures. Ensuring the construction industry follows guidelines for earthquake-resistant buildings, for example, is simple where strong institutions monitor and enforce compliance, and where funds are available to cover any additional costs of safer designs.

Appropriate institutions

Attempts to translate principles of natural disaster hazard management to political crises raise two questions: do we know enough about the hazard, and do appropriate institutions exist to design and implement effective policies?

It is commonly argued that since the most unstable countries are often the least developed, more and better development will prevent conflict. In post-conflict situations, it is suggested that applying conventional development models will achieve political and social reconciliation. However, development cooperation's capacity to yield economic and political progress has been questioned by the disintegration of states, such as Rwanda and Somalia, which once topped the table of aid per capita. Using the strategies of development assistance to improve the quality of relief may well not be the solution: development assistance may contribute to the hazard, not reduce it.

When allocated to regimes guilty of corruption, human rights abuses and promoting inequitable development, aid has contributed to populations' vulnerability. Despite the rhetoric, good governance and human rights conditionalities introduced by many donors in the late 1980s have not been systematically pursued. Many question whether our present pursuit of developmental goals actually improves people's well-being and safety or just enhances the development indicators so often cited as justification for continued investment.

Equating improvement of social and economic indicators with stability risks missing the point that development is inherently turbulent. The aim is not to promote stability at any cost, maintaining an unjust status quo, but to define development's values and objectives, and to ensure that conflict does not become violent.

Achieving this implies a political analysis of the environment in recipient countries: for such an analysis to be operational would mean that development agencies make judgements about the political direction they wish to see their funding support. But for humanitarian agencies, such a potentially partisan approach threatens fundamental principles of neutrality and impartiality.

Linking relief and development raises practical as well as political questions. The division of labour between relief and development is rooted in different agencies' specialised mandates, including those within the UN system. The grey areas between the mandates of the UN High Commissioner for Refugees (UNHCR) and the UN Development Programme, for example, in mitigating the impact of refugees on host communities is a case in point.

It is encouraging that donors are scrutinising their divisions between relief and development. New mechanisms improve internal cooperation and external effectiveness. For example, the European Community (EC) and bilateral agencies, such as the US Agency for International Development (USAID), have used relief and development budgets to create special funds for post-conflict rehabilitation, a task often lost between the relief and development desks.

Unlike relief assistance, development cooperation still has political and economic conditionalities. Making relief more developmental implies decisions about the legitimacy of recipient institutions, deciding who should be empowered politically and financially. If handled insensitively, this could jeopardise the neutrality of relief. Used well, it allows relief to become more sensitive to the diverse and long-term needs of the disaster victims.

Taken to its extreme conclusion, integrating relief and development would need fundamental reform of the international aid system, directed with clear political vision. For many humanitarian agencies, concerned to maintain their neutrality and independence, such a vision of the future is profoundly disturbing. That said, recognition of the essentially political nature of the challenges facing the aid system and recipient countries is giving rise to the second trend of integration: enhancing links between aid policy and foreign policy.

Aid has always been a political instrument, but it was fairly blunt while both superpowers maintained an interest in promoting its humanitarian function. With superpower politics out of the way, other policy concerns come into play. Domestic spending in donor countries is squeezed, creating a need to justify aid budgets. Many see a growing commercial logic of aid: certainly, aid flows are becoming more provincial, with European donors, such as Germany, looking to help the former Eastern bloc, and Japan focusing on South-East Asia. For example, the European Community's Humanitarian Office allocates over 40 per cent of its aid to the former Yugoslavia, even though it represents a much smaller proportion of the global number of conflict-affected populations.

Analysts also point to the use of humanitarian aid to prevent major migration into Western countries. Providing humanitarian aid *in situ* can contain large population movements within national borders, reducing the impact on neighbouring countries. Recent experience in the former Yugosla-

Box 5.2 Time to professionalise the professionals

Effective relief intervention needs the right people available in the right place to do the job. An unprecedented study, co-sponsored by Save the Children Fund, the International Health Exchange, the Register of Engineers and the British Red Cross, and funded by the UK Overseas Development Administration, assessed the problems facing relief workers and their employers.

It concluded that many relief NGOs are poor employers: their recruitment and management procedures are often casual and do not sufficiently support staff working in complex and dangerous environments. Employment practices need to be changed to enable relief workers to develop their skills, and to cope with what can be traumatic experiences.

The report urges relief agencies to develop a code of practice for human resource management so staff receive good support before, during and after field assignments. It proposes the creation of a professional body for relief workers to set, monitor and promote standards for employers and practitioners.

At present, the initiative is focused on the UK and a steering group has been established to take forward the report's recommendations. The potential for international collaboration in the establishment of professional standards is obvious. ■

Figure 5.1 **The system: The increasingly complex international relief system**
A complex web of relationships links the disaster victim with the external financing that allows international assistance to flow. The increasing complexity and interdependence of this system has led many agencies to re-examine how they preserve their independence of action.

Source: ODI, London

via is interpreted by some to represent an example of "enlightened self-interest" by European countries.

Aid is emerging from its cloak of political neutrality, recognising that social and economic objectives will not be achieved unless the political climate is right in recipient countries.

In the 1980s, politics and aid became explicitly linked in recipient countries because of conditionality about human rights and democratisation. In the mid-1990s, some donor countries are integrating ministries responsible for aid policy with those responsible for foreign policy. In Sweden and the United States, for example, such restructuring is the focus of vigorous discussion. This lifts the veil which separated the political function of promoting domestic interests overseas from the supposedly technocratic process of economic and social development.

The pace and quality of this integration has been insufficient to halt claims that humanitarian aid is often delivered as a substitute for clear political direction about conflict and genocide. Critics argue that while humanitarian aid is being politicised, by being used to promote images of caring donor governments, it is neither politically informed nor politically responsible.

Maintaining the neutrality of humanitarian aid requires constant attention. It is a resource which can be used by powerful forces to consolidate their

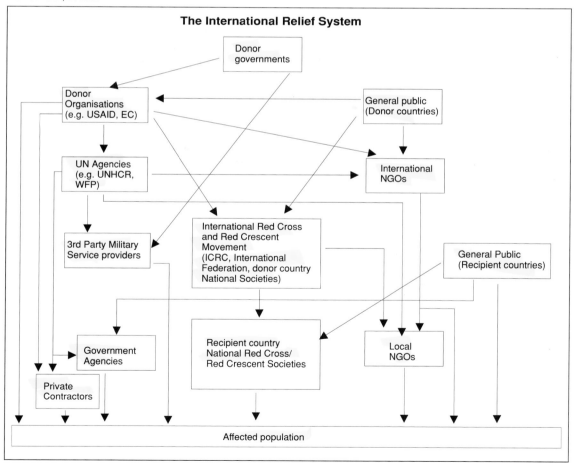

The International Relief System

positions, and thus needs scrupulous practice and careful monitoring. Ironically, politicians frequently rely upon such neutrality to defend policy in- action. In both Bosnia and Rwanda, political intervention has developed much more slowly than the humanitarian response.

Integration of foreign policy and humanitarian action has been further complicated by the use of third-party military forces in complex emergencies (see Box 7.2). In addition to peace-keeping and peace-enforcement within such diverse frameworks as the UN, NATO and the Economic Community of West African States, soldiers are providing humanitarian assistance. This blurring of boundaries threatens the ability of the International Red Cross and Red Crescent Movement, non-governmental organisations (NGOs) and UN humanitarian agencies to defend their claims to be neutral.

This leads potentially to the worst of both worlds: politics which threaten the neutrality of aid but lack any coherence. Developing political coherence to respond to major human rights abuses, not just to the humanitarian fall-out, faces formidable obstacles. Primary of these are the difficulties of building consensus between the powerful member states of such multilateral institutions as the UN and the EC. Even if consensus were achieved by the five permanent members of the UN Security Council, or the European Union Council of Ministers, who would be responsible for ensuring that these powerful bodies were accountable to the people who matter – those suffering the effects of violence?

Contract culture

What does emerge from this debate is a clear need to preserve the ability of humanitarian agencies to provide assistance and protection to all those in acute need, now and in the future. Humanitarian aid essentially addresses the effects of crisis. It is necessary action but not sufficient. Aid workers look to those in the political arena to address the causes of crisis and ensure the humanitarian space for relief to be provided in an impartial and neutral fashion.

NGOs have been one of the primary instruments of the humanitarian revolution. The contract culture, whereby some NGOs have become the implementing arms of donor bodies, has meant dramatic increases in NGO numbers. By virtue of their position on the front-line of relief policy and practice, they have also been the subject of complaints of lack of accountability, poor professional standards and opportunism.

All NGOs are not the same. Alongside the unprofessional maverick are the competent and committed. Standards, to scrutinise NGO activity, to

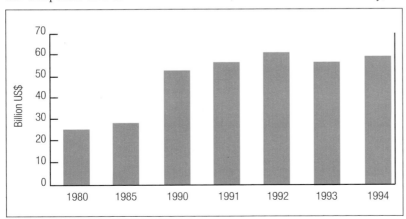

Figure 5.2 **Development assistance: Zero growth in total development assistance**
Total development assistance from the OECD Development Assistance Committee members to developing countries has not increased substantially in this decade, yet during this time many developing countries have undergone radical change and a rapid downturn in their ability to care for the most vulnerable citizens.

Source: DAC/OECD, Paris

promote the good and throw out the bad, are urgently needed. A number of strategies are emerging to confront the problem. NGOs are beginning to develop mechanisms for self-regulation and standard-setting. The global *Code of Conduct for the International Red Cross and Red Crescent Movement and NGOs in Disaster Relief*, now publicly subscribed to by more than 70 organisations and welcomed by 142 governments, provides one example. In the UK, a group of NGOs commissioned a study to explore professional standards and the management of relief workers, proposed the establishment of a system of individual and agency accreditation (see Box 5.2).

As financial providers, donors have a responsibility to monitor the quality of NGO interventions they support. Since donors also finance the UN, standards and a system to guarantee the quality of UN agencies are required. Even if donors improve monitoring and evaluation of implementing partners, it is unclear who will monitor the donors.

While the international community places increasing emphasis on its right to intervene, no parallel rights have been developing to ensure that conflict-affected populations receive effective and appropriate assistance. While these rights exist, at least in law, for refugee populations, there is no similar legislation to protect the interests of the internally displaced, for example.

The international relief system has no capacity to monitor the quality of service provision. As well as setting standards, mechanisms to enforce them are essential. This cannot be done merely by each agency strengthening its monitoring and evaluation systems, since these are often confidential, subject to internal vested interests, and tend to reinforce a pattern of accountability from implementing agency to donor, not to aid beneficiaries.

That pattern reflects the funding of the international relief system, in which the key components form a complex web of inter-relationships, with bilateral donors and the EC playing the primary role in "feeding" the system with funds. These funding institutions, particularly those of the countries in the Organisation for Economic Co-operation and Development (OECD), which provide the vast majority of overseas development assistance (ODA), have witnessed great changes in the past few years.

In 1993, total ODA from OECD countries fell from their 1992 high of 60.6 billion US dollars ($) to $56 billion, recovering in 1994 to $59 billion at current prices. While total ODA has been declining, relief-aid expenditure has been increasing, both in absolute terms and as a proportion of total ODA. Between 1992 and 1993, for example, relief flows increased by nearly 25 per cent. While the rate of increase slowed between 1993 and 1994, relief budgets continued to grow.

Figure 5.3 **Humanitarian spending: Is the bubble fit for bursting?**
Funding flowing into international humanitarian response rocketed in the early 1990s as the constraints of the Cold War were lifted. 1994 was a record year for cash flows, but initial estimates for 1995 and forecasts for the future are less bright. The humanitarian boom may well be over, requiring agencies to focus on increases in efficiency and market share rather than simple growth.

Source: DAC/OECD, Paris

While there is a clear trend for relief budgets to increase, less clear is why they are increasing. Is this a real policy change or a knee-jerk reaction to crisis? Does the rise in emergency expenditure reflect increasing needs, changing public priorities or politicians' preferences for media-friendly relief rather than the slow and risky processes of long-term development?

It might be argued that poor countries, particularly in Africa, are less conducive to long-term investment, reflecting their deteriorating economic and security environments. The spending shift also reflects donor imposition of stricter conditionality: development budgets are subject to political and economic conditionalities, emergency aid expenditures are less so. In Malawi and Kenya, for example, development assistance has been suspended intermittently by some donors, but humanitarian assistance has always continued.

Tracking systems

Of equal concern should be the question: Where is the money going? Given the scale of relief expenditure – nearly $3.5 billion in 1994 – the weakness of financial data is remarkable and should warrant considerable concern in the international donor community. In the absence of strong and rigorous financial tracking systems, relief spending accountability remains relatively poor, and it is difficult to assess accurately trends in relief finance.

The Office of Foreign Disasters Assistance (OFDA) of USAID once maintained the most extensive global relief expenditure database but abandoned it two years ago in a cost-cutting exercise. As the relief industry has boomed, so the capacity for monitoring its finances has declined. Such poor levels of financial accountability would not be tolerated by any major corporation, and it is alarming that it should be acceptable when public and private funds are used to assist vulnerable populations.

In the absence of OFDA data, the OECD Development Assistance Committee (DAC) provides the best alternative. However, donor data provided to DAC are difficult to interpret, particularly because of the allocation of food aid between emergency and non-emergency aid. If food-aid data were included, the proportion of ODA allocated to relief would rise substantially, particularly for the major food-aid donors such as the USA and the EC.

Making more accurate information available means collecting better aggregated data of total donor relief spending. Systems to track resource flows through the complex trail of sub-contracts would also enhance account-

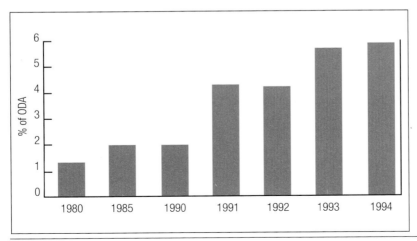

Figure 5.4 **Funding emergencies: The shift from development to humanitarian funding**
Humanitarian funding still only accounts for less than 6 per cent of total assistance spending, excluding food aid costs, but this is up from 2 per cent a decade ago. There is concern that this signifies a shift in donor thinking and a greater willingness to use humanitarian assistance as a foreign policy tool and often in isolation from any long-term strategy.

Source: DAC/OECD, Paris

Sources, references, further information

Griffin, K. "Foreign Aid after the Cold War", *Development and Change*, No. 22, pp. 645-685, 1991.

Institute of Development Studies. *Institute of Development Studies Bulletin*, Vol. 25, no. 4, 1994. See especially Duffield, M., "Complex Emergencies and the Crisis of Developmentalism", pp. 37-45.

Macnair, R. "Room for improvement: Management and support of relief and development workers", *Relief and Rehabilitation Network*, Paper no. 10, Overseas Development Institute, London, 1995.

Macrae, J. and Zwi, A., eds. *War and Hunger: Rethinking international responses to complex emergencies.* London and New Jersey: Zed Press, 1994.

Organisation for Economic Co-operation and Development. *Efforts and Policies of the Members of the Development Assistance Committee.* OECD, Paris, 1995.

UN Department of Humanitarian Affairs. *Aid Under Fire: Redefining relief and development assistance in unstable situations.* UNDHA, with Overseas Development Institute and ActionAid, Geneva, 1995.

ability. The growing number and diversity of agencies involved in the provision and financing of relief has made this both more difficult and urgent.

The number of NGOs involved in relief aid has increased markedly in recent years, reflecting the trend since the 1980s to privatise official development assistance. The expansion in financing has been driven by a greater willingness of the international community to engage within conflict zones, rather than on their periphery. In this context of market expansion, a "contract culture" has emerged which extends the sub-contracting of work downwards and outwards from donors through the UN and the EC, and ultimately to numerous NGOs.

In Goma in 1994, for example, there were over 100 NGOs operating in support of refugees, while in Rwanda itself, numbers reached over 200. In addition to making it extremely difficult to follow the audit trail of sub-contracts, the number of NGOs makes it hard to monitor the standards of different agencies and to develop effective coordination strategies.

Increasing military involvement in relief operations is raising new issues in the allocation and costing of military-humanitarian interventions for ODA accounting. From Bosnia to Rwanda, military forces have undertaken activities ranging from conventional peace-keeping to delivery of medical services. Varied donor practices in accounting for military contributions to humanitarian operations may attribute costs to government budgets for defence, development cooperation or foreign affairs.

As costs rise, relief aid is coming under closer scrutiny. But mechanisms to monitor the international relief system's finances remain poorly developed in tracking both the scale of aid and its complex flow from source to beneficiary.

Interpreting financial trends requires more than counting dollars: evaluating the effectiveness and efficiency of relief spending implies a clear definition of expected outputs and outcomes. Yet the objectives of humanitarian intervention have become more complex, opaque and confused since the end of the Cold War era. It is important to place the financial flows within the broader framework of aid and conflict management in the 1990s.

This chapter talks of "trends": this is perhaps a misnomer, implying consistency, direction and predictability. Subject to an alarming degree of whim, international responses to humanitarian crises differ according to their location, intensity, duration and visibility, and the domestic situation of donor countries. The state of the international relief system reflects a paradoxical trend of globalisation – images of Goma beamed instantly into our homes – and a retreat to parochialism in an environment of profound political and economic uncertainty.

The trend or theme which emerges from a review of the "system" is a growing awareness that much is wrong. The rate of growth of the relief industry in the early 1990s is unlikely to be sustained. Less clear are what reforms of the relief system in particular, and of aid policy in general, will be enacted.

There is great scope for incremental changes at the margins to enhance the effectiveness of emergency aid, but controlling humanitarian challenges will demand more than institutional reform of the international relief system. Far more profound changes in international economic and political relations will be required to protect human rights and to address the causes of conflict. ∎

System crash: Even the most affluent and sophisticated of societies has difficulty coping with major disasters. Reliance on costly technology rather than people, machines rather than people's capacities, can leave developed societies – especially their crowded cities – vulnerable to catastrophe. After Kobe's earthquake, it was the resilience and initiatives of the affected families in the face of cold weather, water shortages and poor communications that provided the basis for the initial response, not the pre-planned emergency systems.

Makeshift kitchen, Kobe, Japan, 1995. Philip Jones Griffiths/Magnum

Earthquake perceptions and survival

The January 1995 earthquake that hit Kobe in Japan highlighted the disaster strength and weakness of a modern city. Strength in the speed and thoroughness of reconstruction and rehabilitation, weakness in the city's dependency upon technological systems.

When the earthquake hit, mass transport systems ground to a halt, the overloaded telephone system crashed, and water mains and power cables were cut. Relief efforts in the paralysed city were slow, patchy and poorly targeted until these lifelines were restored.

Above all, Kobe's lessons are about perceptions not technology. Few in Kobe believed there was a risk of earthquakes or took precautionary measures. The city authorities believed residents would follow pre-designated evacuation plans; they did not. Hospitals believed they were self-sufficient; they were not. As one survivor put it: "The earthquake has taught us not to rely on the administration for everything. Self-help and self-preparedness are the keys to survival."

Finally, Kobe demonstrates how reliant relief efforts are on timely and accurate information. When information systems failed, as phones went dead and people could not get to work, relief efforts were delayed. International information systems, principally television, gave the impression of the city's catastrophic collapse and reinforced previous misconceptions about Japan and Japanese society. This led to much inappropriate international assistance being offered.

The Great Hanshin-Awaji earthquake – as seismologists know the Kobe earthquake – struck at 5h46 local time on 17 January 1995. It lasted 11 seconds, killed 5,501 people, destroyed 106,000 houses and made at least 319,000 people temporarily homeless. Two weeks later it was reported that 4,500 people had lost their jobs as small businesses went into liquidation. The final bill for reconstruction and rehabilitation was estimated at 95 billion US dollars ($) in a country regularly affected by earthquakes, committed to preparedness and with the capital and technological power to back up the commitment. But preparedness is more than the application of technology. Rather than sophisticated technology or careful engineering, this chapter is the story of the people of Kobe and their perceptions of disaster.

Despite consistent warnings from seismic experts, people in Kobe believed that it was an area with a very low risk of earthquakes. This century Kobe suffered a small earthquake in 1916, another nearby killed 1,000 people in

1927 and a third in Totonia near Kobe killed 1,000 in 1943. The most recent earthquake to affect Kobe City was in 1948 when 4,000 people died in the Fukui area in an earthquake with its epicentre well off-shore. Many believed that this fault might slip again so, when the 1995 earthquake struck, local people assumed it was the off-shore fault and that therefore the devastation covered a wide area, rather than being localised around Kobe. This was one reason why there was initially little attempt to ask unaffected hospitals and other services in adjacent cities for help.

Prior to the 1940s, Kobe's citizens had the same attitude to earthquakes as those of Tokyo today: a healthy respect and fear. But such respect requires constant reminders. With no earthquakes in living memory, Kobe's citizens allowed their perception of risk to drift.

Lack of recent earthquakes in a previously active area suggested that tectonic pressures were building up, yet the rate of earthquake insurance among Kobe's households was less than half the national average.

People's perception of risk was also shaped by the reality of recent experience. Kobe suffers typhoons and fires every year. To guard against being trapped in fires, old people are advised to sleep on the ground floor so that they can better escape. Younger people tend to sleep upstairs. When houses collapsed in the earthquake, those sleeping on the ground floor were more commonly killed than those on the upper floors. More old people than young died, guarding against one perceived risk only to put themselves in danger of another.

Most injuries were caused directly by furniture moving during the shaking or by people being cut from broken glass as they tried to escape. Many injuries could have been prevented by fixing furniture in place. Furniture fixing is a key disaster-prevention message promoted in Japan.

In the Tokyo area, where people have a very high awareness of earthquake risks, some 20 per cent of households and offices have their furniture attached to the walls, and televisions and computers strapped to desks. In Kobe, only 2 per cent of households had taken such measures. Since the earthquake, the figure has risen to 30 per cent.

People in Kobe perceived little risk, until the earthquake struck. Those outside Kobe relied on television for their understanding. National television coverage played a vital role alerting national authorities to the disaster's scale and urgency.

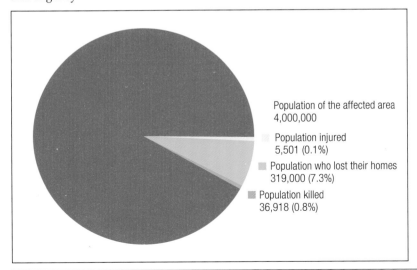

Population of the affected area
4,000,000

Population injured
5,501 (0.1%)

Population who lost their homes
319,000 (7.3%)

Population killed
36,918 (0.8%)

Figure 6.1 **Deaths, injuries and losses**: **Human losses in the Kobe earthquake**
Nearly 10 per cent of the population was directly hit; had their houses destroyed, were injured or were killed. Add to this those who lost their jobs as businesses went under and those who suffered stress and grief at the loss of loved ones and the enormity of the disaster becomes more clear.

Source: Hyogo prefecture administration, Japan

Managers at the local Japan Broadcasting Corporation (NHK) TV station in Kobe believe the television reported as fast and as well as it could. The first TV news bulletin to Kobe and the surrounding area went out at 5h50, four minutes after the earthquake.

At 6h00 it was relayed and broadcast to Tokyo, and at 6h50 footage shot inside the NHK office as the earthquake took place was broadcast. The first live broadcast from outside the NHK building was at 7h23.

Television stations on the edge of the earthquake – in slightly affected areas – filled the initial information gap with pictures and commentary, but showed only light damage, giving a false impression of the scale of the earthquake.

At first, reaction from the rest of Japan was slow. The Japanese damage scale used to report earthquake intensity goes from one (minor) to seven (most severe). Levels one to six are determined by the amount of ground movement, which can be reported instantly from seismological instruments. Level seven is defined as an earthquake that destroys 30 per cent of the traditional wooden houses in the affected area. Three months before Kobe there had been a level six earthquake in Hokkaido which caused no deaths.

Box 6.1 Survivors' stories – 'I was totally trapped'

Mr and Mrs Shinto used to live in a small single-storey house in one of the quieter districts of Kobe. He is 80 years old, she is 70.

Mr Shinto was in bed when the earthquake struck. "I remember waking up and thinking at first that this was just another small earthquake, then the wall behind my bed collapsed on to me. Luckily our water boiler is next to the bed and it stopped the wall falling right on me. Even so, I was totally trapped. I knew my wife was up. I called to her but got no reply and was afraid she was dead."

Mrs Shinto had gone to use the lavatory at the back of the house. "I felt the ground move, sort of round and round, and the plaster from the ceiling came down. I had to kick the door to get out. Our house had collapsed. It had actually moved down the street a few metres.

"I called for my husband but there was no reply. I could see the corner of the bed sticking out of the rubble. Then I heard his voice. Some of my neighbours came round and together we pulled him out, but it took us two hours. The survivors were all out in the street, hugging one another and crying with relief. Most people were barefoot as they had escaped straight from their beds, and this was mid-winter so it was really cold.

"Our son lives nearby. He managed to get through to us later that morning and we went to stay with him, but then there was a gas leak near his house and we had to go to my daughter's in Osaka.

"After a few weeks we moved into this temporary house – it's OK, but small. It is right out on the edge of the city. It's a 20-minute walk to the nearest metro station from here and then half an hour by metro into town. We don't really need to get to town but it is really tough for the younger people who have to go to work."

Mr and Mrs Asano came to Kobe from the countryside many years ago, so they did not have relatives to help them. "I remember the noise of all the plates and glass in the kitchen falling down and the walls came in all around me," recalls Mrs Asano. "I yelled for help and a neighbour heard me but it took five hours to dig us out.

"We managed to get into the local kindergarten for shelter. It was locked and dark but I had found a flashlight somewhere and we got in through a window and took shelter. Other people followed us and it sort of became an unofficial shelter. We were there for three months. I'd no idea what we were supposed to do but someone told me we had to register with the authorities. In April we were assigned to this temporary house. We will apply for one of the new flats the city is building, but there must be a long waiting list.

"Some good things have come of this earthquake. Families are much closer now and so are communities. For those first few months we all helped each other, that's how we survived. No one ever believes it will happen to them. There are simple things you can do like bolting your furniture down and knowing where your valuables, like cheque books and passports, are and always having some water in the house. Water was the real problem after the earthquake." ∎

A month before there had been a level six earthquake which had caused three deaths. Kobe's earthquake was first reported as level six, not an unusual phenomenon in Japan. It was three days, after damage assessment, before it was officially upgraded to level seven.

But by mid-morning, aerial TV coverage made the scale of the devastation apparent to all. The city administration geared up for relief work, and the Japanese Self-Defence Force and Japanese Red Cross Society (JRCS) medical teams arrived.

Volunteer action

In a reaction not seen before in Japan, volunteer relief workers flooded in. About 630,000 volunteers worked in the Kobe area in the first month. Explanations for this phenomenon included instant and nationwide coverage of the disaster, and the fact that it occurred during the student holidays (volunteer numbers fell rapidly as the academic term began).

Again, it was people's perception that counted. They saw devastation on TV, thought they saw a mismatch between what the authorities were doing and the needs, and they had the opportunity to help.

It is a truism that earthquakes are only disasters if there are vulnerable people in their way, but the nature of vulnerability can be fickle.

If the earthquake had struck two hours later, people would have been awake and dressed, and the death toll in the wooden houses might have been much lower. Four hours later, with the mass transport systems full of commuters, car-choked roads and offices filling up with workers, the death and injury rates could have been much higher.

So what actually happened when the earthquake struck?

Professor Murosaki of Kobe University's architecture department explains: "The movement of the ground in the Kobe earthquake went way beyond all the limits our engineering had planned for. Our highway bridges are built to withstand a ground acceleration of 3.5 metres/second/second. In the Kobe earthquake the ground accelerated up to 8.17 metres/second/second.

"Normally the ground moves a total of maybe 20 cm during an earthquake. At Kobe it moved up to four metres.

Box 6.2 Health care – four phases and an earthquake

Doctors in Kobe identified four phases in post-earthquake medical assistance showing how survivors and the medical system coped and gradually returned to normal.

Phase 1: First two or three hours. Most rescues took place. People buried under the rubble either pulled themselves out or were pulled out by other survivors.

Phase 2: First 48 hours. Systematic, assisted rescue starts. On-the-spot first aid treatment is given. Emergency evacuation to the hospitals still functioning starts for the most serious injuries.

The most important is on-site triage to give priority to the most urgent cases. Transport of the seriously injured to hospital is difficult and rare.

Phase 3: Day three to two weeks. Basic relief deliveries start: food, clothes, water. Medical staff visit evacuation shelters administering first aid and treating patients suffering from mild trauma. Acute diseases brought on by winter weather and poor living conditions – colds, gastric ulcers – treated in temporary medical centres. Systems to supply drugs and dressings start to function.

Phase 4: Week three to two months. Chronic diseases, such as hypertension and diabetes mellitus, need regular treatment. Patients start to be treated through the usual medical system rather than temporary centres. Normal long-term welfare services are established and counselling, to help people rebuild their lives, begins. ∎

"We normally expect lots of horizontal shaking and design buildings to withstand this. Here we had great vertical *and* horizontal movement, as well as a twisting effect, all at the same time. None of our engineering models ever looked at such a situation before and so we had no idea how buildings and other structures were going to react."

The earthquake was a "direct hit", with no warning before the big shock. Off-shore earthquakes usually give two or three seconds of tremors before the big shock, enough for people to dive under tables, grab a torch or a radio. Many people took up to an hour to get out of their houses as their way was blocked by fallen furniture, the time meant it was dark and the electricity was knocked out.

Professor Murosaki adds: "We also had problems with many of the bridges. We live on alluvial land, sand and soil deposits next to a chain of mountains. Many bridges had their end supports firmly fixed to the mountain bedrock but the middle support was on the alluvial deposits in the valley. The alluvial material became liquid, just like wet sand on the beach, and the central supports moved.

"The worst problems were with the two-storey wooden houses that most people live in. These are where most of the deaths occurred. Many of these homes collapsed. Wooden construction is traditional in Japan and, if you look at the temples and houses that are more than 100 years old, you can see how clever the old building style was. The wooden frames of these buildings have joints which are not fixed permanently but which will allow movement, like fixing the joints together with an elastic band rather than metal nails.

"The technique required highly skilled carpentry. About 100 years ago we started using European wooden building techniques with braces and tie bars which aim to create a rigid structure. Such a structure is quicker to build and requires less skill and, for all but the most severe of earthquakes, it works fine. But, of course, this was a severe earthquake. We can't go back to the old style of wooden building, the carpentry skills no longer exist and I don't know that you could automate the process to allow for the amount of building we need today.

"But there is something seriously wrong with the way we approach earthquake engineering. About 100 universities in Japan teach earthquake engineering and architecture, but I know of only one or two that teach anything about wooden house design. Yet 80 per cent of Japanese people live in wooden houses. We teach our students how to build office blocks and

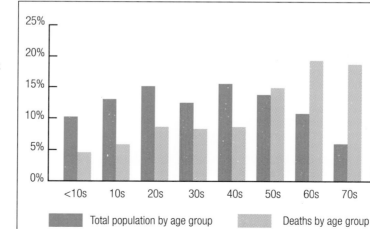

Figure 6.2 **Disaster vulnerability: Disasters seek out those most at risk**
By comparing the percentage of Kobe's population in each age group with the percentage of the dead from each age group, it can be seen that older people were at greatest risk of death. Earthquakes, like all disasters, seek out the vulnerable.

Source: Japanese Red Cross Society

hotels, where the prestige and money is. Maybe we will see some change now, but I'm not hopeful."

Earthquakes need vulnerable people to become disasters, and some people are more vulnerable than others. Like most of Japan, Kobe has an ageing population, but the elderly were disproportionately killed. The JRCS Kobe branch has a programme where hospital nurses visit some 40 elderly, house-bound people every day. The earthquake killed 14, one-third.

Those worst off before the disaster suffered the most. Many of those still left in temporary housing are the very elderly and disabled. They suffer additionally because the temporary housing is not designed for disabled people and few have health facilities or day clinics for the elderly nearby.

Foreign workers

The earthquake also disproportionately affected Kobe's 130,000 foreign and migrant workers. Most work in low-paid jobs for small businesses, many of which closed after the earthquake, while those with limited Japanese did not know how to obtain help.

For the survivors of the earthquake the priorities were to find shelter from the winter cold, to get in touch with loved ones, to find water to drink in a city where all the piped water had failed and, finally, to find food.

"We had a plan with some 360 evacuation centres identified. Schools, gymnasiums, that sort of place. We intended to register all those who needed assistance through these centres," recalls Kasuo Ikawa of the Kobe City administration, "but when the earthquake hit, up to 600 places were used as evacuation centres. People did not wait to be told which were the official centres, they sought refuge where they could. We found it really difficult to find out where they were and so it took ages to get help to them. It is even worse for those who went to stay with relatives but still need help and are entitled to assistance with re-housing, for instance."

According to one relief worker, the problem was exacerbated by the cramped conditions common in Japanese cities: "There was hardly any open ground at first to set up relief posts, nowhere to move the debris to. We needed open space to sort the debris into wood for burning, steel for reusing and concrete for disposal in land fill. To top it all, there was nowhere to dispose of garbage."

Despite the problems, within a week evacuation centres were functioning effectively and work had begun on constructing 40,000 temporary houses in parks and on the city outskirts. Rebuilding Kobe may take up to ten years. The temporary housing is designed to be demolished after two years. Disaster experience suggests that the temporary housing will be inhabited for far longer, even though each person has on average only 26 square metres of space. Better-off people have started rebuilding their homes, so those still living in temporary housing tend to be the poorest. The temporary housing is far from people's former homes, work places and children's schools – and it is not earthquake-proof.

Children have another problem. Japanese cities often lack open space, and many playgrounds have been used for relief work and temporary shelters. Volunteer agencies, such as the YMCA, stepped in with day-care centres and play-leaders.

Faith in Japanese technology was shaken by the earthquake. The bullet train stopped working, as did the electricity and telecommunications networks. The disaster-information system, designed to ensure that the city administration could act rapidly on good quality facts and figures, did not work well. Modern cities rely on systems and lifeline connections to keep them alive. The earthquake showed just how vulnerable systems, particularly centralised systems, can be to disaster.

The normal disaster-reporting system moves up from the local police and fire services to their municipal offices, which then report to the Prefecture office and hence up to the National Land Ministry, which in turn informs the Prime Minister's office.

Both the police and fire services have their own satellite telephone communications systems which were still working after the disaster, but information did not flow because the local officers were so overwhelmed with relief work that they had little time to make reports.

Local government currently relies on the Disaster Prevention Administration (DPA) radio for public information communication. Nearly half of Japan's municipalities now have this system, but in the most damaged areas of Hyogo Prefecture where Kobe is situated, the system had not then been fully installed.

Some municipalities had mobile units to collect disaster information but only one had installed a public address system to convey emergency messages.

Before the DPA radio, local government in this area had installed a $1 million emergency system which relied on satellite telephone links. It proved useless as roof-mounted satellite dishes were knocked out of alignment, computers and fax machines were thrown off desks and smashed, the water-cooled back-up power generation system failed because the water system was broken and the final back-up batteries ran out of power after five hours.

The administration had to rely on its older short-wave and VHF radios. They worked but, as Kasuo Ikawa of the Kobe City administration explains, human factors slowed the relief effort.

"Our initial relief action was slow. We identified four reasons. The first I have told you about, we didn't know where people were, which shelters they were in. Before that, we faced a problem with our own staff being victims and many of those who were not victims had to help pull their relatives and friends out of the rubble, before making their way into the office through a city with no transportation. This is a city reliant upon mass transport systems: metro, train and bus. Without them and with the streets blocked, everyone had to walk. Journeys which normally took half an hour ended up talking half a day."

These problems were further exacerbated by the lack of connectivity between the communication systems of the different services and administration. Each service had its own short-wave communications system on separate frequencies. Links between services and communication with municipal offices was supposed to take place via the telephone system, but that did not work.

Of the 1.44 million telephone lines in the area, 285,000 were out of action because of damaged switch gear and a further 193,000 because the lines themselves were cut. More of a problem was the overload caused by people trying to call into and out of the Kobe area. In the first day there was a 50-fold increase in the number of calls, effectively paralysing the system.

Figure 6.3 **System crash: City support systems affected by the earthquake**
Kobe, like all modern cities, depends on systems to function, yet these very lifelines were destroyed in the first few seconds of the disaster. Designing disaster-proof utility systems for cities is now a major challenge for urban planners.

Source: Hyogo prefecture administration, Japan

System	Damage	Date repaired
Electricity	1 million consumers cut off	23 January
Gas	845,000 consumers cut off	11 April
Water	1.27 million consumers cut off	17 April
Sewage	260 km of pipes broken	End April
Telephone	478,000 lines cut	End January

Public pay-phones had another problem. Much of Kobe uses phone-cards, but telephones using phone-cards need electricity. Even where phone lines were working, the lack of electricity left many public phones useless.

Back-up power generators should have been the emergency solution, but again a system failure occurred. Hospitals and many services did have back-up generators but these were designed to provide power for a few minutes, not a few days, so they could not cope. Many were water-cooled, but the water supply had been cut, so they overheated and shut down. While 97.8 per cent of hospitals were out of action, many would have operated if their service systems, especially electricity, had worked. The lesson has been learned: Tokyo metropolitan authority is replacing all its water-cooled generators with air-cooled versions.

Radio and TV have a vital role in disaster response in Japan and are required by law to cooperate in broadcasting public information. They also operate a public service allowing people to have personal messages broadcast to reassure relatives and friends. In the aftermath of the earthquake, TV stations received 50,000 message requests but could only broadcast 35,000. In previous disasters affecting a few hundred thousand people the system worked, but four million people were in Kobe's disaster area and the system broke down.

During the first few hours of the disaster, helicopters flown by TV crews caused real problems when their noise made it very difficult to hear the faint cries of trapped people. Because each helicopter was from a different station it was difficult to contact them all and request periods of no flying.

Earthquake disasters bring their own particular brand of injuries and medical needs. When modern, urban structures collapse, cement dust kills by suffocation most of those trapped in the first 15 minutes. Trapped survivors often suffer from crush syndrome, an accumulation of toxic substances in body tissue. When released, the toxic substances can cause death through kidney failure.

Kobe's victims had a very old age structure, older than that which is the norm in Japan. Many people were dependent upon regular medical help and medication to treat chronic diseases and ailments. These chronic conditions quickly became acute after the earthquake when drug supplies and treatment were interrupted.

The chief doctor at the Kobe Red Cross hospital remembers the day of the earthquake. "Ours is one of 69 hospitals in the Kobe area. After the earthquake, staff tried to get into the hospital knowing that they would be needed, but they were victims too. Thirty-two per cent of our staff lost their homes in the earthquake. By 6h00 the first patients were coming in by themselves,

Sources, references, further information

Christine, Brian. "Disaster Management: Lessons Learned." *Risk Management*, Vol. 42, no. 10, 1995.

Overseas Development Administration. *Megacities: reducing vulnerability to natural disasters*. London: Thomas Telford, 1995.

United Nations Department of Humanitarian Affairs. *The Great Hanshin-Awaji (Kobe) Earthquake in Japan: On site relief and international response*. Geneva, 1995.

Japanese Red Cross Society: see Chapter 14.

Kobe on WWW:
http://pele.nando.net/newsroom/jsources.html

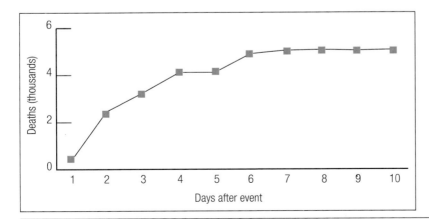

Figure 6.4 **Reporting deaths: Reporting the dead, or the presumed dead?**
In Japan, death toll figures are based on proven deaths, not estimated as in many other countries. Day-by-day reports therefore gave the impression that people were continuing to die in large numbers, yet in reality this was merely a result of the reporting system.

Source: Hyogo prefecture administration, Japan

mostly with superficial wounds. By 7h00 patients were being brought in by their relatives and, by midday, ambulances were bringing people faster than we could unload them.

"We were lucky. Hospitals in Japan are run by universities, by municipalities and by a few private companies. The Red Cross is the only real network of hospitals, with some 92 across the country. When the earthquake hit, our sister hospitals did not wait to be asked or told to send help. We had a system in place because of our specific role in the country's disaster-preparedness plans, so we knew how to best utilise incoming help. Other hospitals were not so fortunate. They tended to operate in isolation, had no network to back them up and did not have a system ready to integrate help when it did come."

According to Professor Haruo Hayashi of the research centre for disaster-reduction systems at Kyoto University, Kobe's reliance upon technical fix-it systems is the norm across Japan. Disaster-preparedness and -response systems have been based on physical engineering. Kobe showed that not enough attention had been paid to the people side of preparedness.

Box 6.3 Foreign perceptions and international response

The reactions of foreign governments and aid agencies and their offers of assistance were governed by perceptions of Japan and the scale of Kobe's tragedy. Four factors combined to give the impression of a major disaster beyond the coping capacity of Japan.

First, most Westerners have little knowledge of Japanese geography. There was an implicit assumption that Kobe – 500 kilometres from the capital Tokyo – was actually nearby and thus that Tokyo was also badly affected.

Second, unlike other countries, Japan's system of reporting deaths in an earthquake is precise, using not estimates but the actual body count. The death toll rose steadily in the first few days, giving the impression that people continued to die from lack of relief efforts.

Third, historical recollections of the 1923 Tokyo earthquake, which destroyed the city and affected a wide area, led to assumptions that Kobe was a similar event. The Kobe earthquake affected a very restricted area, the main damage being confined to a strip one kilometre wide by 20 kilometres long.

Finally, international TV coverage of the disaster tended to concentrate upon the sensational pictures and the stories of tragedy and need, reinforcing the false notion people already had of earthquakes in Japan.

Japan had great difficulty refusing foreign assistance. Swiss sniffer dogs arrived: they are usually used to find people lost in a wide area, but in Kobe people were trapped in their small houses and most sites were already identified by neighbours and relatives. The offer was well meant, but technically was not an effective use of resources.

Most people trapped in earthquakes are rescued in the first few hours; few survive beyond two days. European and American rescue teams had an impossibly short window of opportunity. However, Japanese rescue workers and survivors saw an unexpected value in foreign rescue teams. The sense of solidarity, of a shared tragedy in which all countries were concerned for Kobe's people, had a marked effect on local morale.

The disaster was a battle ground for other issues. One American non-governmental organisation (NGO) heard that many survivors had the 'flu and sent a 747 jet full of treatment for colds.

Osaka, next to Kobe, is a centre for drug production, the drug was labelled in English not Japanese and bringing in pharmaceuticals broke import regulations, so the authorities refused to distribute the goods. The refusal was interpreted as an example of obstructive Japanese bureaucracy and unfair trading practices.

Two companies offered hundreds of mobile phones. Phones alone are of no use, needing licences, numbers and channels. When the local authorities pointed out that they were short of staff and asked for help rather than just donations of hand sets, this was again cited it as an example of obstructive bureaucracy.

Two French medical NGOs arrived. One, seeing little for them to usefully do, retreated to Tokyo and worked with their Japanese fundraising branch. The other insisted on taking a high profile in the area, but interpretation staff had to be assigned to work with their doctors, who did not speak Japanese, thus pulling skilled and much-needed staff away from more urgent tasks. ∎

"In the past, few would admit that there was a need for psychological help for disaster victims and relief workers, but our research has clearly shown that around half of the earthquake victims suffered from post-traumatic stress syndrome. Many of them need counselling and a few need psychiatric care. We have no system to deal with the counselling. Those people will either go to an expensive and possibly inappropriate psychiatrist, or get nothing."

The story of Kobe is one of perceptions and attitudes. The lessons are in how to counter these misconceptions. Since few believe they will be caught in a major disaster, constant reminders are needed to make local disaster preparedness work.

Too often we build survival systems which depend on shaking foundations, from telephone lines to electric power grids and mains water supplies. There may be no alternative to such an approach for normal times, but in disaster-prone areas back-up systems have to be more self-sufficient.

Box 6.4 From cash to telecoms – national response and

The Japanese government assigned essential disaster-relief and -preparedness tasks to major relief organisations. The Japanese Red Cross Society (JRCS) was one of the organisations charged with identifying the most vulnerable people in the disaster area and, in cooperation and coordination with central and local government authorities, carrying out the following missions:
● to train and make available at least five medical relief teams recruited from the 47 JRCS chapters;
● to coordinate medical and midwifery services with other medical services;
● to send out medical relief teams to provide medical treatment, midwifery services and the handling of corpses;
● to mobilise Red Cross volunteers to assist relief operations, and to organise and train local volunteers at the community level;
● to collect contributions in cash and in kind to be distributed to disaster victims; and
● to establish Red Cross relief volunteer corps at city and village level.

Red Cross action

The earthquake left Kobe's telecommunications systems severely damaged and the local Hyogo chapter of the JRCS faced enormous difficulties in setting up relief activities. But a neighbouring JRCS chapter in Okayama was able to establish radio contact with Hyogo and forwarded information to JRCS headquarters, who immediately sent a staff member to Kobe to contact local JRCS members.

The day after the earthquake, 18 January 1995, the JRCS had set up a disaster-relief task force at its headquarters to consolidate relief activities, as well as a local task force in Kobe to expedite the distribution of supplies and medical teams. In all, more than 1,500 JRCS staff were involved in the Kobe operation.

Mobile medical teams

On the day of the earthquake, the JRCS assembled 23 mobile medical teams from in and around the Hyogo prefecture. Each team consisted of six people – doctors, nurses and a medical coordinator – whose job was to provide emergency medical treatment for injured and sick people at the evacuation centres.

On the same day the JCRS set up ten medical centres from which the medical teams could operate and, on 21 January, upgraded them into central relief centres. At peak there were 12 offices in the most devastated areas of Kobe, Nishinomiya and Ashiya, and 20 to 30 teams in operation full-time. Over a period of ten weeks, 5,965 staff members in 979 mobile medical teams treated more than 38,000 people.

Medical services at Red Cross hospitals

The 126-bed Kobe Red Cross Hospital in the city centre initially took a leading role in providing medical services to the affected area and, in January, dealt with over 400 patients a day, over 200 of which were disaster-related. The hospital was at one time so overloaded with injured people that local Red Cross meeting rooms in the hospital building were converted into temporary wards.

Doctors, nurses and pharmacists from neighbouring Red Cross hospitals, as well as volunteers, assisted the hospital. Injured patients who required

---

Finally, disaster-information systems need to become two-way channels or, as Kobe showed, the best plans can be rendered useless as people rightly take their survival into their own hands. In major urban disasters there must be a way to get useful information to the citizen so that they can take action themselves, rather than waiting for the city authorities to help. ∎

the relief tasks of the Japanese Red Cross Society

special treatment were transferred to the Red Cross hospitals in Himeji, Osaka, Takatsuki and Otsu. The Tokyo-based Science and Technology Agency lent the hospital the only mobile CAT-scanner available in Japan to aid emergency treatment.

In February, the hospital treated, on average, 300 patients daily. One-third of them had been affected by the disaster.

Mental-health services

To treat mental-health and stress-related problems, the JRCS dispatched a study team and launched an investigation at ten relief stations, while 21 Red Cross hospitals opened hotlines for mental-health counselling for the disaster victims.

Blood centre

Power and water failures in Kobe City temporarily forced the Hyogo Red Cross Blood Centre to suspend normal operations.

The Red Cross Blood Centres in Osaka and O-kayama cities launched a nationwide blood donation campaign, which helped prevent any shortages of blood products in Kobe or elsewhere. The Hyogo Red Cross Blood Centre resumed full operations in February.

Distribution of relief supplies

In cooperation with the Kobe municipal government's task force, the JRCS transported and distributed emergency relief supplies from its headquarters and chapters, including relief goods donated by domestic firms.

Volunteer action

Many volunteers were involved in the Kobe relief operation, including JRCS youth and student groups, trained disaster-relief volunteers, ham radio operators.

The Red Cross Flying Corps, composed of volunteer amateur pilots and ground crews, operated 67 small aircraft and three helicopters to fly medical teams, relief goods, blood and other supplies and personnel.

Tracing service

The JRCS headquarters and the Hyogo chapter accepted tracing inquiries from overseas, and by 27 March had received 1,787 inquiries and successfully traced 1,577 people, both Japanese and foreign nationals.

Fundraising appeal

The Japanese Red Cross Society launched a nationwide fundraising campaign for disaster victims.

In three months, it raised 100 billion Yen ($1 billion). The distribution committee in Hyogo prefecture includes representatives of the Hyogo prefectural government, Kobe City municipal government, Japan Broadcasting Corporation (NHK), and the JRCS Hyogo chapter.

The first distribution targeted the families of those killed or missing, or whose houses had been destroyed. By the end of 1995, at least 75 per cent of funds had been distributed, with a second distribution targeting the seriously injured and children who lost their parents. ∎

Crisis contradictions: Dilemmas abound for aid agencies in almost every disaster, but the Rwandan crisis seems to have more than its fair share. There are issues of justice and assistance, local needs versus those of refugees and displaced people, security and development. For the Rwandan people, whether in detention centres, refugee camps or their own villages, life has to go on, but for those responding to disaster, resolving the contradictions is crucial if lessons are to be learned and future work to be improved.

Displaced people's camp, Rwanda, 1995. Sebastiao Salgado/Magnum

Contradictions in crisis

Rwanda has been a defining disaster for the international humanitarian community, forcing a new scale of response and speed of reaction, and bringing with it, as the *World Disasters Report 1995* and the *World Disasters Report Special Focus on Rwanda* investigated, complex dilemmas about politics, human rights and humanitarian priorities.

Unlike Bosnia or Somalia and to a lesser extent Mozambique, Angola and other complex emergencies, the crisis of Rwanda has unfortunately not been such a defining disaster for the other players on the stage – governments, the military, the media – whose attention and resources have been concentrated elsewhere following the first immediate emergency period.

In part because of this lack of serious attention, for over two years Rwanda's dilemmas have been explored but not resolved, the crisis inside and outside the country has continued, and the future of Rwanda and its people – and the people and nations of the Great Lakes region, especially those suffering amid the simmering civil war of Burundi – remains unpredictable and precarious.

Violence or its threat, forced repatriations, reduced living standards and loud political rhetoric for domestic consumption fail to find solutions for those caught up in Rwanda's disaster.

They deepen the crisis, especially for the humanitarian agencies doing their job with inadequate resources, high outside expectations and limited political action to address the causes of the crisis.

Yet two years or more of refugee camps – refugee cities of wood and mud, plastic sheeting and corrugated iron – and the grinding muddle of Rwanda trying to rebuild, have offered lessons for those agencies, and perhaps for the other players. These lessons are best understood in the context of the contradictions that Rwanda forces upon those who would try to help the most vulnerable, and thus the constraints imposed on the way agencies plan and act.

Written during the chaotic change in and around Rwanda in early 1996, this chapter will explore those contradictions and examine any lessons being learned, though the ever-shifting situation does not encourage firm predictions.

As with much of the *World Disasters Report*, its perspective takes full advantage of the unique position of the International Red Cross and Red Crescent Movement, active in every country of the region at all times as both international agencies and indigenous National Red Cross Societies, working in partnership with every government, all parts of the United Nations (UN) and most non-governmental organisations (NGOs), and community groups.

Even more than most refugee disasters, Rwanda has from the start forced upon aid agencies the contradiction of planning for today and for years to come. Today, because more than two million refugees have immediate and pressing needs; for years, because the trend in recent years – from Afghanistan to Liberia, Central America and Viet Nam – has been for refugees to stay for years in their host countries as efforts are made to resolve the issues that led to their flight.

In past refugee disasters, those forced to flee have been offered long-term asylum, international protection and substantial support. Today it is very different. The shift has involved less international donor funds and political will to support large and long refugee operations, and far greater hostility from host governments to refugees, including efforts to first block the arrival of refugees and then encourage their return.

Box 7.1 Evaluating Rwanda – the verdict, $1.4 billion later

A $1 million evaluation of the Rwanda emergency reported in early 1996 with an assessment that looks likely to set the agenda for future debate between donors and aid agencies.

Led by Denmark and funded by a donor consortium, the Joint Evaluation of Emergency Assistance to Rwanda used five teams to examine the historical background, attempts at peacemaking and peacekeeping, the $1.4 billion humanitarian response, reconciliation and reconstruction, and to prepare a synthesis report.

The synthesis report suggests that a complex of factors, some based in Rwanda's history and others more proximate, contributed to the genocide, and there were significant signs that forces in Rwanda were preparing the climate and structures for genocide and political assassinations. People both in the region and the broader international community ignored, or misinterpreted, the significance of these signs. For those who were planning genocide, this indicated an unwillingness to intervene. Key actors in the international community thus share responsibility for the fact that the genocide was allowed to begin.

According to the report, the international community failed to stop or stem the genocide, by hesitating to respond and vacillation in providing and equipping peacekeeping forces. In this regard, it shares responsibility for the extent of the genocide.

The essential failures of the response of the international community to the genocide in Rwanda were and are political. Since key political issues have yet to be resolved, the crisis continues, as does the necessity for massive allocation of humanitarian resources.

As the extent of flight of people from Rwanda became clear, the international humanitarian assistance system launched an impressive and, on the whole, effective relief operation. The international response saved many lives and mitigated large-scale suffering. Improved contingency planning and coordination, increased preparedness measures and adoption of more cost-effective interventions could have saved even more lives, as well as relief resources.

The report suggests that several factors shape Rwanda's options for recovery, such as the overt rearming and reorganisation of the former leadership, military and militia in Zaire, which have posed a threat of war for well over a year.

While some donors have been forthcoming, the international community's failure to provide adequate support for Rwanda's government undermines future stability and development efforts. Insufficient attention and resources have been given to survivors of genocide and war inside Rwanda.

An essential element of reconstruction in Rwanda must be establishing an effective system of justice through which perpetrators of genocide are punished, thwarting the "culture of impunity".

Property, land-use rights and other requirements for successful economic and social integration of pre-1994 refugees weigh heavily on the government.

The report concludes that lasting resolution of Rwandan political problems will be achieved only in the context of both an inclusive political system that reflects the underlying principles of the Arusha Accords, and the crisis across the Great Lakes Region, especially in Burundi, where politically motivated violence has created an explosive situation that threatens regional security. ∎

In part it has simply to do with the cost of such operations, but there is also the sense that without the Cold War, there is less motivation for the West in particular to assist those affected, from Afghanistan to Viet Nam or Cuba, and pressures have grown for refugees to go home.

An assumption that refugees will stay – and the political context of the host country to allow that assumption – immediately offers enormous benefits to both refugees and aid agencies. For refugees it can mean the opportunity to work or farm, important not just to cut the costs of refugee relief but in enhancing personal dignity, security and self-reliance. For aid agencies assisting host governments, it offers the chance to plan effective services in health and education.

But Rwanda has been different, with pressure almost from the start for refugees to return. The reaction of Zaire and Burundi has been symptomatic of the change in global context for people on the move. For agencies and Rwanda's refugees, the disinterest and hostility has been both traumatic and costly. Unable to plan in confidence, agency contingency planning – stockpiles, extra staff, maintaining additional reserves, constant short-term actions and high cost options – cannot have improved the quality of work and may have cost millions of dollars more than necessary.

Money, mandate, machinery

Amid wild fluctuations of conflict and politics, Rwanda has seen the contradiction for agencies of simultaneously working both with and without state structures, and inside and outside the country's borders.

In a series of emergencies in recent years, aid agencies have learned, often unwillingly, to work with little or no government structures. In the 1980s and 1990s, but especially since 1989, the power and importance of many governments has declined as free trade, liberalisation and privatisation promote the private sector and uncontrolled markets as the guarantees of individual freedoms. Alongside this, many states have simply collapsed, leaving little structure to work with.

The negative impact is clear to see, since planning in a humanitarian wasteland is often costly and pointless, while the potential positive dimension – forcing agencies to be directly responsive to the needs and capacities of beneficiaries – has been far less in evidence.

Disintegrating states, such as Somalia and Liberia, offered nothing for international agencies to work with. Even in countries with governments, the lack of cash or legitimacy, or both, has often left little with which to work in partnership; for example, no functioning ministry of health or credible regional councils, a health system in slow collapse, hospitals without drugs, bandages or doctors, local health workers without aspirin.

The early stages of Rwanda's crisis saw the curious reversal of the state destroyed within the country while much of its former personnel moved *en masse* into the refugee camps around the country. Entire communities, with their elected or traditional leaders, arrived and resumed life in a camp, while nothing functioning was left behind. Senior politicians and businessmen moved out to regional capital cities or Europe, while their workers, army, police and professional staff drove or walked into Zaire and Tanzania, where many began planning reinvasion.

Aid agencies found the human infrastructure of the former state living in the camps, with armed militias enforcing their continuing rule, preventing refugee return through propaganda and intimidation. Again the contradictions arose. Agencies were aware that the old structures had allowed or fostered the slaughter, but relief systems, such as food distribution, initially used the previous, hierarchical leadership – *commune, préfecture* – rather than trying to reach the community-level *cellules* or even individual families.

Those working with refugees have operated with that reality for more than two years, while their colleagues inside Rwanda first worked with no state at all, then with a new state that lacked all infrastructure, and finally with a state that is internationally recognised but lacks much of the aid necessary for it to function.

In 1996, the Rwandan state exists and is expected to live up to international standards of human rights and justice, if only for the practical reason that if it does not, few refugees will return. Committed to bringing the guilty to justice, it keeps tens of thousands in jail without trial because it cannot find enough money to gather evidence, pay lawyers and judges, and stage fair hearings within any reasonable time frame, let alone provide adequate food and health care for prisoners.

The pledging conference for Rwanda in early 1995 brought promises of 707.3 million US dollars ($). By July 1995 less than 10 per cent had arrived, and most of that has been channelled through aid agencies rather than the government. The United Nations Development Programme estimated that by the end of 1995, pledges had risen to just over $1 billion, but still only half the funds initially pledged had been disbursed to help survivors of the slaughter. This contrasts with the $1.4 billion already spent on refugees outside the country, including those responsible for organising the killings.

Box 7.2 The military in Rwanda's disaster – should the

Many foreign military forces were involved in the international community's response to Rwanda's crisis, undertaking security duties, relief roles or a mixture of both, and working within a range of mandates. The experience of Rwanda raises many major questions about whether military forces should ever be used in humanitarian operations except in a peacekeeping capacity.

The Joint Evaluation of Emergency Assistance to Rwanda assessed the military in three ways:
1. How much military contingents increased the security of threatened populations and created "humanitarian space" for effective relief assistance to be provided.
2. Performance in providing relief assistance and support to relief agencies.
3. Lessons for future involvement in humanitarian operations.

In summary, the evaluation suggests that when troops were ordered to protect civilians, military forces were effective, saved many lives – though far fewer on occasion than they claimed – and by creating humanitarian space, assisted the relief effort.

This implies that in Rwanda military forces should have been used far earlier, in greater numbers and to far greater effect, particularly in stopping genocide, limiting the humanitarian emergency and thus reducing the substantial cost in lives and resources.

The achievements were far less consistent when troops did not have a security mandate, concentrated on their own security or focused on relief activities.

All military forces probably undertook or were drawn into some relief activities or support for agency relief efforts. This ranged from airport management to burying bodies. Some units carried out some functions – from water tankering to road repair – extremely well.

According to the Joint Evaluation, the provision of relief work and support for aid agencies was uneven, since soldiers have little experience of humanitarian roles and most units were unprepared because their governments had spent months refusing to commit them to Rwanda in any peacekeeping capacity.

Military contingents arrived and left on political timetables that often had little relation to needs; crucial commitments, such as the delivery of heavy machinery for site work, were not fulfilled; and much military equipment – though crucially not the US water-treatment plant in Goma – was often not ideal for a mass relief operations.

When troops did not have a mandate to stop civilians being killed, there was the bizarre sight of heavily armed troops in helmets and flak jackets leaving meetings about relief tasks to get back to their bases before nightfall at a time when hundreds of civilians were being killed every week and

The Kigali government is frustrated by the failure – as it sees it – of the international community to offer aid or fulfil its promises of support and furious at donor talk of aid "conditionality" about human rights, for example, when no such conditions were considered for refugees whose ranks include mass murderers. Suspicious that agencies are exploiting the "emergency" to maintain their income, Rwanda has resorted to taxing donor-backed aid agencies for vehicles, radio and imported supplies, normally exempt from duties, in an attempt to get some money from governments.

Of course, the former rebel movement that is now the government in Kigali has its own contradictions that agencies must understand and live within: it is internationally recognised but has no national democratic mandate; it must govern, set taxes and impose laws, yet has little political or economic infrastructure and thus limited real power, and has problems in efficiently absorbing and using aid; and it must meet the needs of those inside the country now without prejudicing the rights – especially in land and property – of those yet to return.

Meanwhile, the Rwandan government has faced conflict with aid agencies grown used to operating without any government controls or state structures in so-called failed states and to resisting what they perceive as growing interference in their work that could threaten their independence. The gov-

soldiers get off the humanitarian front line?

aid workers in T-shirts walked around refugee camps unprotected.

The evaluation outlined three major issues of concern: cost, stand-by capacity and civilian-military cooperation in such complex emergencies.

On costs, it found that the same activities cost considerably more if undertaken by a military force than civilian contractors or aid agencies. For example, military airlifts cost four to eight times the commercial alternatives and were in some instances far less flexible. When military forces used aircraft for supplies even though ground routes were open, the costs were 20 to 40 times more expensive than commercial trucking. In many cases, military costs in emergencies are billed to the government's humanitarian aid budget.

When the humanitarian aid system was perceived to be overstretched in the early stages of Rwanda's emergency – a perception not always accurate, especially for the commercial sector's airlift capacity – the military was able to offer its stand-by capacity and logistics expertise.

The evaluation suggests that a humanitarian aid stand-by capacity developed in commercial or non-governmental sector would be considerably cheaper than the cost of funding maintenance of such capacity within the military. It would also be predictable, experienced, flexible and available even when the political will for a military operation did not exist.

On cooperation, Rwanda showed a level of confusion on both sides between agencies and military units because of their different mandates and lack of mutual experience. Some agencies may have had issues of principle at stake, since cooperation with the military may be perceived internally and externally as taking sides in conflict.

These issues – and others – were exposed by the UN High Commissioner for Refugee's (UNHCR) crisis measure to ask governments to provide service packages that had not – as service packages should be – been previously agreed and prepared for immediate operation. In many cases, governments turned to their military forces to implement packages ranging from airport management to water supply.

Overall, the military are very expensive; operate under unreliable and unpredictable political control; are usually inexperienced and unprepared for humanitarian missions; have communication problems with aid agencies and among beneficiaries; divert much-needed funds from humanitarian budgets; and prevent aid agencies or commercial contractors from developing the capacity to respond at lower cost.

Finally, in today's world of conflict, such as Rwanda, soldiers are more urgently required to do their prime function: apply politically controlled violence or its threat, to save lives and create humanitarian space. ∎

ernment does not see it that way; they talk of establishing systems, creating structures and planning strategies within which agencies must operate or join those which have already been asked to leave.

Donors, the public, the media, human rights groups – all have their own perceptions of the Rwanda disaster, its "victims" and how they were then treated. This again created contradictory pressures on aid agencies and relief staff.

Guilty victims

Until Rwanda, all refugees were, more or less, regarded as innocent victims caught in the crossfire of crisis, even if in some cases combatants' wives, children and elderly parents were ensured a secure food supply and shelter while their husbands, fathers and sons fought on. But some of Rwanda's "innocent" victims – though how many, no one knew – were guilty of the most appalling crimes and human rights violations.

The theory was and is clear. It is for governments in Africa and elsewhere to determine guilt or innocence, either directly or through the International Tribunal for Rwanda, and for UN human rights machinery, such as the UN Human Rights Commissioner and the UN Centre for Human Rights, to help that process. Under the mandates of most aid agencies, it is not the job of relief workers to accuse, arrest, put on trial or sentence people, even those widely believed to be guilty of evil crimes; it is not their job even to stop feeding them.

Box 7.3 Training for the future

Crisis and conflict may abound in and around the refugee camps in Zaire, Burundi and Tanzania, but people still need to live and build lives. Children are born, midwives and health services are needed. In Benaco camp, Tanzania, the International Federation is training traditional birth attendants (TBAs).

At the graduating ceremony, the men in the room were not only the husbands – one was actually a graduating TBA, receiving his diploma and TBA kit along with 22 Rwandan women. Back home he had been a traditional healer, but he wanted to be of help in other areas as well. The location was the Federation's Referral Hospital in Benaco camp near Ngara, Tanzania, and the group was the second graduated since the courses began in November 1995.

There was pride in the graduate's faces – and those of their spouses – as they accepted diplomas and kits from UNICEF's Area Representative for Tanzania, Agnes Aidoo, visiting the hospital in mid-February.

The graduation ceremony was a joyous affair with traditional Rwandan dancing. The jubilant graduates were of different ages – some very young, some not so young. Their new TBA kits, donated by UNICEF and supplemented by the Federation, include an apron, a plastic sheet for deliveries, hand-towels, soap, cotton wool, eye ointment, napkins, a simple weight, a razor blade for cutting the umbilical cord and other basic items. Benaco now has a total of 78 TBAs, most of whom attended a six-week course run by the Mother and Child Health Unit.

The TBA's are attached to the 34 health information teams run by the Tanzania Red Cross Society (TCRS) in Benaco and Lukole camps. "Many of them have been working as TBAs before," says Bernadetha Kagemulo, TRCS head nurse at the Referral Hospital.

"The purpose of the training is to increase hygiene and the safety of deliveries in the camp. Most women in the camp want to give birth at home, and with a sizeable contingent of TBAs we hope to make life for the women – and their babies, of course – easier and safer."

The TBAs need not worry about idle days – the birth rate at the refugee camps in the Ngara area is around 6 per cent, so that every year around 6,000 babies are born in Benaco camp alone. ∎

Once the dimensions of the slaughter within Rwanda became clear, some believed it was the task of aid agencies to change their role and become policeman, judge and jury. Even within relief agencies, some felt that although the primary responsibility of aid workers is to assist the needy, they should inform on, or at least should not employ, those they suspected of crimes.

On occasion, neutrality and impartiality were dismissed as irrelevancies rather than principles that have allowed the needs of the vulnerable to be met for many years.

By casting aside the humanitarian for the judicial, this view would have required aid agencies to ignore questions of evidence and the presumption of innocence. At a more practical level, if staff were known or even suspected of accusing refugees of crimes, it would put them at appalling risk in camps where Rwandans were hacked to death for merely saying they wanted to return home.

Meanwhile, those whose job it was to investigate and apprehend criminals among would-be refugees – host governments and the UN – failed to move with sufficient speed, even when agencies such as Oxfam UK & Ireland and Save the Children UK tried to shame the UN over its slow progress by meeting some of its budget shortfalls for human rights field officers.

Ready or not

Rwanda has been a series of disasters, from the killings themselves to the Goma arrivals and Tanzania influx, then the cholera epidemic, further massacres and camp deaths from stampedes and militia violence, and the continuing shadow of a disintegrating Burundi. Across the region, other crises threaten, including the collapse of states that would make Rwanda's problems look simple by comparison.

Once again agencies feel pulled in at least two directions. They are expected to react instantly to each new crisis despite the burdens of present work; indeed, there is an expectation that aid agencies, alone, should in some way pre-empt, prevent or at least mitigate the next crisis while still dealing with the last, even though the budget for the last has not been fully met and contingency accounts are all but empty.

Preparedness costs money, money in advance, but the public mainly react to crisis after the event. Instead of taking a long-term view, even one that would almost certainly save money by reacting swiftly and effectively, governments are now reacting to this crisis and others – if they react at all – as if each humanitarian crisis was just another item on the domestic political agenda. But Rwanda was never high enough on anyone's political agenda.

In recent years it has been possible to prevent a disaster through prompt government action – the anti-drought operation in Southern Africa in the early 1990s proved that – but this lesson seems hard to learn and even harder to apply elsewhere, as budgets fall and attention focuses on national issues not global needs. One well-placed UN official likened the position of aid agencies today to firemen asking for donations as they race to a blaze.

Yet in the heat of Rwanda's crisis, when money was available because politicians and their public wanted action, the lack of preparedness, common standards and well-funded agencies with plenty of experience meant millions of dollars were wasted, in using military forces which cost four, eight, 20 or 40 times the agency or commercial rates to move supplies, for example.

Perhaps even more important than money is the political context of the Rwanda operation. As in many other humanitarian emergencies, agencies talk of the political vacuum of their work in Rwanda. They accuse politicians, governments, donors and in some cases the UN – the collective expression of states – of using humanitarians as a cover for their failure to act, while signalling by their inaction that they are unconcerned about the fate of Rwanda and its people.

The political vacuum is the most obvious expression of the broader retreat of government worldwide, a lack of confidence that makes it all too easy for a Rwanda and its millions in need to be moved down the political agenda until it disappears from media and public attention.

As aid spending has fallen and been overtaken in value by private financial flows, as the UN wavers in direction and support, as donor states focus spending on fewer countries, as crises are met by holding actions, with peacekeeping troops, short-term agency actions and the occasional flurry of political meetings, the need for consistent, clear pressure to find political solutions within which humanitarian actions can be most effective becomes ever starker.

Symptoms and causes

Neither soldiers nor aid workers can also be politicians, yet in the political vacuum both military forces and relief agencies have ended up being pushed or pulled onto the Rwanda "stage", feeling as if they are playing someone else's role. Aid agencies in particular believe that they have been denied the resources and humanitarian "space" to do their own job to full effectiveness. The result will prolong the misery, not solve the crisis.

Millions of refugees and displaced people who will not or cannot go home are symptoms of a crisis, as well as being part of a disaster in their own right. The causes of Rwanda's crisis are far more political than they are "ethnic", "tribal" or due to any other victim-blaming euphemism.

The causes are also economic. In 1993 alone, drought brought a 21 per cent fall in the harvest. Between 1989 and 1993, falling world coffee prices, worsening weather, structural adjustment programmes and the impact of the civil war from October 1990, caused the country's GNP per person to collapse by 40 per cent. With inequalities of wealth, that average figure conceals a crash in the living standards of the poorest.

Aid agencies can do little about the politics and economics of the Great Lakes region, not even much – despite support for fair-traded goods – for the price of coffee. If the region is not to become a permanent holding camp for the exiles from a state failed by other states, governments and intergovernmental organisations will have to help Rwandans find the solutions.

Finding ways to deal with those who planned and executed the slaughter will be a key element in those solutions. Destroying the culture of impunity surrounding such mass murder is no part of agency work but will be a precondition to an effective end to humanitarian actions.

Hate radio

In a similar way, agencies have neither the right nor ability to close down Rwanda's powerful "hate radio" broadcasters, but unless those with the duty to deal with criminal activity do act, agencies can only struggle on in an atmosphere of corrosive propaganda or spend money that has other priorities on rival stations.

In 1996, there are still some among UN agencies, donor governments and NGOs – though few among the refugees – who assume that the solution to Rwanda's problems is for everyone to go home and try to live and work together. Some also expect that those organisations funded by donors should cooperate in whatever means are necessary to make this happen, such as by running transit camps when Rwandans are forced to leave countries around Rwanda and walk back.

Caught at the focus of these contradictions is the UN High Commissioner for Refugees (UNHCR). Not only does it have a dual role – meeting refugee needs and ensuring their protection – it is also under the control of the governments of the UN, especially the permanent five of the Security Council. UNHCR also faces issues of major overlap in its functions in disasters with other UN agencies, especially the World Food Programme.

Its protection mandate may mean letting refugees stay where they are; the reality of political pressures may mean "assisting" refugees to return. Its protection mandate means it has to pay Zaire's Presidential Guard to carry out security duties that Tanzania's police and army have done without subsidy; political pressures mean cooperating with Zairean troops as they cut off camps to encourage refugee return.

In the past, when refugees usually arrived because they were oppressed, hungry or poor, improvements in their home country – better government, weather or economics – provided a real logic to a return and a new start. It was rarely the situation that those fleeing were in many cases brutal killers and their active or passive accessories, who would then be asked or pressured to go back and live again in their killing fields side by side with the relatives of those they murdered.

But if the logic of a swift return by Rwandans is flawed, the pressures remain to prevent any new influxes and to squeeze the refugees out.

The average calorie value WFP expects to give refugees is 1,900 kcals a day, higher if they are expected to do more than light work; in early 1996 the ration in Tanzania was 1,900 kcals and in Zaire around 1,540 kcals, though the latter level was expected to rise to 1,700 kcals because of refugee pressure.

In theory, these figures mean a slow death sentence for many; in practice almost two million refugees are still alive through trade, local work, farming and other means, though some of the most vulnerable may be in great need, as malnutrition rates in various camps suggest.

Refugees might feel that the low ration is a major part of efforts to make host countries so unattractive that they are forced to go home. The reality is certainly that the food supplies are low and slow, because of problems in the pipeline donors are unwilling to commit resources, and because agencies today are very conscious that refugees must not be better off than the surrounding population, with whom they may be sharing water, land and fuel.

The contradiction here is that if food rations are deliberately kept low and better targeting systems are not introduced, agencies will have to either accept more malnutrition and deaths or create safety nets of complex and costly additional feeding programmes. In reverse, if most refugees are doing well on starvation rations, then the rations should be reduced further to cut costs and allow better targeting, just so long as refugees are allowed – even encouraged – to be more self-sufficient through work, farming and trade.

Camp violence

Food or no food, the camps around Rwanda, particularly those near Goma, have seen tension between local people and refugees, and violence within the camps. This leads agencies into the contradiction of perceiving a need to improve security but realising that bringing in armed people could actually make matters worse.

As Somalia proved, armed guards hired by agencies do not necessarily stop violence. While they may deter casual robbery, they will have little impact on politically motivated groups or professional armed criminals. Armed guards attract attention, flag valuable supplies or frightened foreigners, suggest that aid agencies favour one violent group over another, and make aid workers a part of the conflict, rather than outside it.

While aid workers may be fearful, those always most at risk in Rwandan refugee camps are other refugees, especially the most vulnerable, the weak, sick, poor, old, young and women. But how should agencies expect to improve their protection when the host countries are reluctant to police camps, and the troops provided may be unreliable, ill-disciplined or sufficiently badly paid to make looting or extortion an attractive option?

Of course, while agencies simply do their job, feeding those in need, they may well know that without outside policing or military control, the camp militias will flourish and be fostered as they plan campaigns of intimidation within the camps and military offensives against the new Rwandan state.

If troops are on the spot, as UN peacekeepers or direct government forces, it seems obvious to use them for their intended purpose – to combat violence or its threat – so the most vulnerable do have some protection and aid staff can help them, rather than have soldiers driving trucks or prescribing medicines at far greater cost than either aid staff or commercial contractors.

There is also the broader question of where the arms used by all sides came from, and which governments either deliberately armed rival groups or failed to prevent their own weapons being sold or resold to such groups.

Hold the line

Rwanda's crisis has offered plenty of contradictions in the multitude of pressures – some of them external, some from within – on aid agencies to lower their standards. The most obvious pressures are political and financial, preventing agencies from doing work they know to be essential or doing such work badly because it will not be well funded by governments unwilling to get involved or publics uninterested because the crisis is being given no political or media priority.

A particular problem for Rwanda is that the disaster has not and will not go away; the commitment by agencies and their funders needs to be substantial and long term. Yet donors are looking for shorter, sharper, cheaper solutions.

The media itself provides another pressure to lower standards, tempting agencies to rush into disasters to maintain income even when their skills are not required, or to do things for coverage rather than meet real needs. The issue here is not the media, but agencies playing to it.

In the short term, there is fact inflation – the numbers always overstated, the needs over-emphasised – in the long term, a temptation to maintain the illusion of crisis when, for most refugees, their personal disasters ended a few weeks after arrival, once they had food, water and shelter.

Competition is good when there are ways of measuring performance. But when neither those giving the money nor those receiving the goods have any means to judge, at least in the short term, competition means waste and duplication, as well as distracting attention from the long-term needs and the local agencies that must be built up to carry out reconstruction and be there for the next disaster.

Freedom versus control

There are contradictions for agencies – they need to have freedom and independence but effective response in complex emergencies needs coordination, even control; enthusiasm and enterprise are invaluable, but professionalism and experience are equally important. Speed and profile are necessary in disasters but consistency and commitment are ultimately more valuable to those in need.

Sources, references, further information

Millwood, David, ed. *The International Response to Conflict and Genocide; Lessons from the Rwanda Experience.* Steering Committee of the Joint Evaluation of Emergency Assistance to Rwanda, Copenhagen, 1996. Text on website: http://www.ing.dk/danida/rwanda.html

Africa Rights. Various publications, 1994-1995. Africa Rights, 11 Marshalsea Road, London SE1 1EP, UK, tel. (44)(171) 717 1224; fax: (44)(171) 717 1240.

Human Rights Watch/Africa and Human Rights Watch/Arms Project. Various publications, 1992-1996. Human Rights Watch, 485 Fifth Avenue, New York, NY 10017-6104, USA, tel.: (1)(212) 972 8400; fax: (1)(212) 972 0905; email: hrwnyc@hrw.org

International Federation of Red Cross and Red Crescent Societies. *World Disasters Report 1995.* Geneva, 1995.

International Federation of Red Cross and Red Crescent Societies. *World Disasters Report Special Focus, Under The Volcanoes: Rwanda's Refugee Crisis.* Geneva, 1994.

US Committee for Refugees. *World Refugee Survey 1995* and other publications. US Committee for Refugees, 1717 Massachusetts Avenue, N.W., Suite 701, Washington DC 20036, USA; tel.: (1)(202) 347 3507; fax: (1)(212) 347 3418.

Agencies did fail to deal correctly and swiftly with cholera and dysentery, but did succeed in getting camps running, food delivered, water available and medical care established. Good, but not good enough.

By contrast, elements often portrayed as the "successes" – arrival of military forces, airlifts, provision of big equipment – concealed many failings. Much took too long to arrive, costly airlifts were not that well coordinated, and military forces were used for tasks commercial contractors and agencies could have done at lower cost, and on the whole were not used for their main purpose – controlling violence or its threat – where they would have done the most good, improving camp security when killings were a daily outrage.

The ultimate contradiction is, of course, that Rwanda's real success story will only happen in the months and years to come, with peace and justice, political stability and economic growth, built consistently and quietly without media coverage or political drama, supported by international agencies and other governments but led and managed by the Rwandans themselves. ∎

Mental wounds: In many disasters, the basics – food, water, shelter – were all that could be offered, and meeting physical wants left psychological needs to be addressed. That is changing fast. In Oklahoma, the psychological impact of the bombing may take decades to heal, but thousands appreciated the expert counselling and assistance available in the days and weeks following the disaster. From Chernobyl to Bosnia, Rwanda to Sri Lanka, the trauma of disasters and the value of psychological support is being recognised.

Victims and relatives, Oklahoma City, USA, 1995. Bruce Davidson/Magnum

Psychological support in disaster response

Disaster response has traditionally included delivery of shelter, food, clothing and medical needs, but a series of crises – from Bosnia to Rwanda – have demonstrated the need and value of psychological support, or disaster mental health, for victims and workers. Focusing on the mental-health response to the 1995 bombing in Oklahoma City, this chapter will address:
- the impact of disasters;
- the need for psychological support;
- integrating psychological support with traditional disaster relief;
- guidelines and resources for development of psychological support programmes;
- the need for collaboration between involved entities; and
- lessons learned which will assist development of future disaster-response capabilities.

On 19 April 1995, at 9h02, Oklahoma City, Oklahoma, was rocked by the largest terrorist act on American soil. The blast from the home-made fertiliser bomb which destroyed the Alfred P. Murrah Federal Building was felt 50 kilometres away, causing extensive damage to ten square blocks of this busy city in the heartland of America. Hundreds of buildings were damaged, including ten that collapsed, and over 500 businesses were interrupted. In all, 169 people died, including 19 children in the second-floor day-care centre of the Federal Building. More than 500 were injured. At least 249 children lost one or both parents, while hundreds of other people lost close relatives or friends.

There have been over 6,500 documented terrorist events worldwide since 1980, resulting in over 5,000 dead and 12,000 injured. Most of the terrorism directed at the United States has occurred outside its borders, so Americans have felt protected from this form of violence. Feelings of complacency and security were swept aside as news of the bombing spread.

The fact that this act of terrorism occurred in a mid-sized city in the heart of the United States, and was carried out by fellow Americans, caused a nationwide increase in anxiety and apprehension. The scenes from this bloody disaster were played repeatedly on television, prompting reactions of horror. Adding to this increased fear were numerous "copycat" bomb threats to schools, hospitals, businesses and governmental agencies across the United States.

The American Red Cross chapter in Oklahoma City is less than ten blocks from the federal building, so Red Cross workers were on site within five minutes of the explosion. They joined medical and community volunteers and the emergency services in initial search and rescue, assisted with triage and first aid as the wounded were brought out, and provided care for the survivors and their family members. The local Red Cross mental-health team, supported by mental-health professionals from the Veterans Administration Hospital, provided care and comfort to the injured and to arriving friends and relatives as they awaited news.

As the day passed and it became obvious that there were few survivors left in the building, the area was sealed off to prevent anyone unauthorised from entering the site and becoming secondary victims. One nurse died from injuries received during rescue efforts. Search and rescue teams from around the country began arriving to support local agencies.

Trained Red Cross disaster workers, recruited through National Headquarters in Falls Church, Virginia, began to arrive from across the country to assist the local volunteer and paid staff, many of whom had been involved without respite for more than 15 hours. A particular challenge and additional workload was dealing with untrained but well-meaning volunteers. Up to 1,000 people a day came forward to volunteer or donate goods and supplies. In addition to providing critical response services, the chapter remained open 24 hours a day for several days to process these volunteers and accept their donations.

As in many disasters, the large number of volunteers demonstrated the overwhelming desire of the community to assist. Response organisations and workers found it difficult to pay for anything within Oklahoma City as businesses and individuals wanted to thank and help those directly involved. This high regard for rescue workers provided additional emotional support.

The Red Cross began providing food services to emergency workers, survivors, family members and others waiting at the site as rescue efforts got underway. Critical Incident Stress Debriefing Teams, associated with the police, first aid and other rescue services, set up a programme to psychologically prepare rescue workers for the task of locating and removing the injured and dead, and to debrief the emergency-response teams at intervals during their work.

Compassion centre

Disaster mental health services were provided via three venues: an overnight shelter, a walk-in "compassion centre", and outreach teams travelling into the community.

A shelter was opened by the Red Cross for those who were displaced from an apartment building and other housing close to the bombing. The shelter also became an assistance centre for family members arriving from other cities to await word on missing relatives. As well as practical help, this was where many took advantage of counselling offered by volunteer clergy and mental-health professionals as information about the recovery efforts filtered in.

Families began to develop a mutual support network, sharing both good news and bad. Mental-health professionals were assigned to each arriving family to provide comfort and support. They helped to verify information, which assisted in suppressing rumours and protected the families from overly intrusive members of the media.

The second site was the compassion centre which was set up within hours of the explosion. The Red Cross, in coordination with the Office of the Medical Examiner, Funeral Director's Association, the clergy, National Guard, federal agencies and other organisations, managed the facility so that information could be gathered, shared and disseminated quickly and accurately. This was where the official death notifications were given to family members.

A team of mental-health professionals was assigned to work with each family as they waited for information. The team allowed individuals to rotate shifts, thus facilitating continuity in support and decreasing the number of new people that family members had to encounter. A day-care area was set up for children whose families were on site. Volunteers whose specialty was paediatric counselling were assigned to this area.

An additional group of workers was set up to support families once information had been received verifying a relative's death. This team included a representative of the Medical Examiner's office, who did the formal

Box 8.1 Gathering global expertise on trauma

The International Federation of Red Cross and Red Crescent Societies' Reference Centre for Psychological Support was established in Copenhagen by the Danish Red Cross in March 1993.

It provides advisory services on psychological support to the International Federation and National Societies, collates information about National Society activities and needs, and is the secretariat for the Federation's Working Group for Psychological Support.

The Danish Red Cross has offered psychological support programmes for many years. In 1990, it gained critical experience providing support to the victims of the Scandinavian Star ferryboat disaster. The Danish Red Cross and the Federation convened the First Consultation on Psychological Support in Copenhagen in 1991.

Established after that consultation, the Working Group on Psychological Support meets twice a year with representatives from the Federation, the Reference Centre and National Societies with significant experience in psychological support, such as the American Red Cross, Belgian Red Cross, British Red Cross, Danish Red Cross and previously the German Red Cross.

The Second Copenhagen Consultation on Psychological Support took place in June 1995, with representatives from 16 National Societies to discuss natural and technological disasters, HIV/AIDS, and worker and delegate support.

The Working Group's tasks and output includes:
● a range of materials, including guidelines for psychological support and steps to develop a psychological support programme;
● a set of recommendations to the General Assembly of the Red Cross in order that the International Red Cross and Red Crescent Movement as a whole might advance;

● a survey of a broad group of National Societies to learn about their experience in psychological aid, their interest in its pursuit and their specific needs; and
● the newsletter *Coping With Crisis*.

The guidelines for the steps to develop a psychological support programme were written with the 169 members of the Federation in mind, but contain the elements that any humanitarian organisation could use to begin work in this field:
1. Determine interest level and priority in relation to other programmes.
2. Appoint a coordinator with overall responsibility for moving forward.
3. Identify unmet psychological support needs in the community and in National Societies.
4. Get advice and materials from the Federation Reference Centre.
5. Include volunteers and mental-health professionals in the group to plan programmes.
6. Determine the programme's scope and identify the human and financial resources required.
7. Set priorities to identify the most important or most basic pieces to put in place first.
8. Identify groups to train as staff and groups to target with psychological support.
9. Detail policies, guidelines, roles, training and relationship to existing programmes/agencies.
10. Provide for supervision of psychological support workers by mental-health professionals.
11. Prepare documents for training workers and materials for those affected by disaster.
12. Implement the programme.
13. Publicise the programme through the National Society and the community.
14. Evaluate and modify the programme as a result of your community's evolving experience and the global experience collated at the Reference Centre. ■

notification, a member of the clergy and a psychologist or psychiatrist with a specialty in bereavement or trauma. The notification area was isolated from the rest of the building to ensure privacy as families began the grief process.

Another group at the compassion centre was assigned the task of debriefing each of the teams at the end of their shifts. Debriefings were also offered for all workers assigned from the various agencies on site. Initially, few took advantage of the services, but as time went on and the number of death notifications increased, the demand also increased to the point that sessions were held hourly from early morning until late at night.

The third area for service delivery was outreach teams, which made hospital visits or went to clients' homes to meet individual needs. Each team included a Red Cross family service worker to assist with any material needs, a nurse to offer guidance and assistance relating to medical or burial needs, and a mental-health professional for psychological support.

Rescue efforts

Shock at the bombing was felt throughout the country. Rescue and recovery efforts were constantly on television. Red Cross chapters and mental-health professionals across the country began to receive calls from children, parents and teachers. Some children were afraid that their parents would not come back from work or that a sibling would be killed at their day-care centre. Parents and teachers wanted to know how to help their children.

A national 24-hour no-charge telephone line was developed to give special support to the emotional needs of children. In five days, approximately 100 volunteer mental-health professionals responded to hundreds of calls from 37 states. Just over half the calls were from adults inquiring about children; the rest were from children themselves.

At every service delivery site, disaster workers, both volunteer and paid staff, were strongly affected by this disaster. Many developed strong "ownership" of the response, with a negative impact on them personally and on their ability to function at their usual high levels.

Typical behaviour was refusing to take breaks, leave at the end of a shift or take days off. Many rescue workers worked 14 hours a day or longer, although there was more than adequate help available to keep to eight-hour shifts. Some chose to sleep on site rather than go home or to hotels. This decreased their ability to make decisions, manage their personnel, or downsize the operation when that became appropriate.

Many workers suffered mood swings, increased anger and sadness, frustration and guilt. Relationships suffered. Most workers experienced intrusive thoughts and dreams, inability to concentrate, anxiety and sadness – all normal reactions to a highly abnormal event.

Disaster mental health – helping normal people have normal reactions to an abnormal event – is a relatively new specialty for professionals. Mental-health services are usually limited to:
- crisis intervention;
- identification of stress or issues which cause survivors to have intense emotional reactions or which may prevent them from gaining control over their life and recovery effort;
- education; and
- problem-solving, including identifying positive coping strategies.

For most survivors of natural-trigger disasters, these techniques will assist the individual in a rapid return to a pre-disaster state of functioning. Many studies have shown that there are differences in how an individual and community react to "man-made" disasters.

So-called natural disasters, arising from earthquakes or landslides for instance, have been described as having a clearly defined "low point", after which recovery can begin. Technological or man-made disasters are often more difficult to recover from, with frequent uncertainty about how or when the incident began, when it will end, and what continuing effects survivors may face. Acts of terrorism prevent a psychological "low point", as survivors may remain fearful of a recurrence. Survivors may feel that the event is never truly over, and will have difficulty beginning or completing the recovery process.

Survivors of man-made or technological disasters can have varying reactions, based on whether the particular disaster was caused by an act of commission or omission. Research has identified more anger and frustration among victims of acts of omission. In this situation, it may not be possible to identify an individual or group responsible; victims may blame the government for not having adequately protected them.

The terrorism of the PanAm crash at Lockerbie, World Trade Center bombing in New York, and the Oklahoma City explosion were essentially acts of commission. Each event also raised questions regarding the failure to enact a security plan to prevent or warn against such occurrences. For many, lingering fears and anger were turned towards those in government who "failed to protect".

Relief workers found that few in Oklahoma expressed anger at the alleged perpetrators. Instead, anger was directed at family members, or governmental and relief agencies. Many mental-health professionals felt that this misdirected anger was due to the inability of individuals to accept that one person could have attacked so many innocent victims. It was easier to deny that reality and turn sadness into anger directed elsewhere. Mental-health workers had to give plenty of support and counselling to those who faced this anger.

Belief system

Many factors may affect how an individual reacts to a traumatic event. The most important individual factor is the level of personal loss, and the meaning of that loss. Death of a spouse or child will have a significantly greater meaning than loss of a home, career or personal possessions. Prior history of trauma, such as having survived the loss of another relative, previous disaster losses, job loss, a recent move or unresolved grief from other events may have an impact on disaster recovery. Vital is a strong sense of self-esteem, a belief system to enhance an individual's ability to survive stresses in life and a strong support system. This may exist through co-workers, family, friends, religious affiliations or community agencies.

The most commonly shared reaction to disasters of any size, and the first element toward possible crisis, is stress. Stress is the body's reaction to demands made on it. It affects people physically, mentally and emotionally. All disasters will cause some level of stress in both disaster survivors and workers. This is usually expressed in increased levels of anxiety, tension, apprehension and fear. Time, education and supportive counselling will usually dispel these emotions, allowing the affected individual to move towards recovery.

A certain degree of stress can also be positive. Good stress, also called "eustress" ("eu" as in euphoria), can lead to motivation. The more flexible and extroverted we become, the more we turn stress into eustress. Bad stress happens when a need for perfectionism turns stress into distress. In disasters, adhering to this perfectionism when one is trying to create order out of chaos, is what helps put an individual into crisis.

What makes one incident a crisis and another a trauma for survivors? Trauma usually has a sudden, life-threatening quality that stress does not. A trauma is an event that makes you realise you could have died; makes you feel powerless, helpless or overwhelmed; is unexpected and often a time of great fear.

Initial reactions to terrorism may vary from shock, disbelief and numbness, to anger, horror, fear and helplessness. Over time, reactions can turn to further anger, depression, isolation, withdrawal, loss of interest in life, emptiness, guilt, increased use of alcohol and drugs, and domestic violence. Acts of terrorism are clearly traumatic. Psychological reactions are far more serious, prolonged and difficult to recover from without outside support.

It is not uncommon for rescue workers to suffer inordinately due to a traumatic event's psychological impact. Some workers may walk away from careers and families while others may try to make similar big decisions while in an emotionally unstable state. Children express fear of returning to school, and parents may be reluctant to return to work or be apart from their children. Those awaiting information on the fate of missing relatives, especially children, will bounce between persistent and inconsolable grief, and hope and anxiety.

Many survivors and relatives of victims may develop symptoms of post-traumatic stress response, including sleep disturbances, frequently recurring, intrusive thoughts, dreams, or recollections of the event and how their loved ones were killed or injured. There may be a persistent avoidance of situations which remind the individual of the event, feelings of detachment or estrangement from others, numbness and unresolved grief. These reactions can also occur among disaster workers. Continuation of these symptoms, which may disrupt their ability to function normally in a social or work setting may be signals of Post-Traumatic Stress Disorder. Counselling and/or medical treatment may be required.

In areas where there is a strong religious influence, residents frequently look to their local clergy for support and guidance during times of crisis. Many who believe that prayer will help them through any life crisis may seek professional mental-health support to help them cope. That, in turn, causes a high level of stress in the clergy – for which they should seek professional help – as they attempt to find answers that may provide some level of comfort to their congregations.

Ripple effect

How do you identify those who may need psychological services and plan to meet their needs? To decrease the intensity of reactions and long-term psychological impact of terrorism or other disasters, early identification of victims and early intervention are vital. Need can be identified by what could be considered the "ripple effect", expanding as time passes.

Those with a relative who died or was seriously injured may need longer and deeper support. Those with missing relatives may need one-to-one support during the wait. Residents displaced due to the disaster will need assistance next.

The acquaintances of those killed or injured generally need services. Included in this group could be members of churches, social and work groups. Clergy and hospital staff, who are usually closely involved in assisting survivors, are frequently overlooked. Residents who will see daily reminders of the disaster may be affected. In industrialised countries, the impact of media coverage may affect people countrywide. Support will be needed by those involved in the disaster rescue, recovery or relief efforts.

In Oklahoma City, unlike many other disasters where many individuals refused mental-health support, a wide range of usually independent individuals came forward to request help, including professionals from the police, fire and medical services, and the clergy.

Mental health is one component among many needing to be fully integrated into a disaster plan. As well as identifying the target population, a disaster mental-health plan of action is needed. How are mental-health services provided in the area, and how receptive are the residents and the primary disaster responders to mental-health support? Is anyone responsible for the community's psychological well-being? What are the special skills and experiences of any private practitioners, and who else is looked to for psychological support?

Resource guide

In the context of disaster-response in North America, part of preparedness planning involves the development of a resource guide with contacts for leaders in the mental-health community. Skills and knowledge particularly necessary after any mass casualty event or terrorism include: experience and training in psychological assessment, trauma, bereavement and loss, critical incident debriefing, and crisis intervention. A roster of individuals with these skills will ensure appropriate and timely disaster response. A 24-hour crisis phone line is essential. Many individuals may be afraid to personally present their symptoms, but willing to share their feelings over the phone. Assessment and referral can be provided immediately for those in distress. The phone line should be working within 24 hours of an incident.

Throughout the planning process it should be clear that disaster mental-health services must be very concrete. Victims are focusing on dealing with their physical and material losses. Frequently, helping access resources and providing information will reduce stress. Psychological support for both victims and workers must be integrated into other activities of disaster recovery.

Several important lessons were learned while attempting to meet the diverse needs of the American people after the Oklahoma City explosion.

A staffing centre for mental-health workers should be set up as quickly as possible. Volunteer counsellors must bring appropriate documentation of their credentials. Volunteers should be screened to assess their motivation and any special skills. Immediately begin assigning volunteers to specific shifts, usually no more than eight hours, and preferably no more than four hours if they are working in an intense emotional environment. Keep accurate records of all volunteer contact details, credentials, skills and schedules. Use respected leaders in the local mental-health community to assist with screening and supervision of volunteers.

Emotional debriefing

Assignments should be based on skills and education. For example, those with a background in grief, loss or bereavement should be assigned to teams that will work directly with those who have lost a family member. Those with backgrounds in crisis intervention or trauma should be assigned to teams supporting those who were injured, had family members injured or those experiencing strong emotional response. All workers should also be given an opportunity for emotional debriefing during their work and at the end of their assignment.

All rescue and assisting workers, and volunteers, should be assigned specific shifts as soon as possible and must be required to leave the job site at the end of their shift. Debriefing may assist them with the transition from the disaster environment to their home. Education about possible emotional reactions, with suggested coping strategies to assist with relaxation and sleep, may help. Mandatory days off should be scheduled at least twice a week during the initial high impact stage following the disaster when stress is highest. Staff should be rotated out of the assignment after no more than 14 days and prevented from taking further stressful disaster assignments for at least 30 days.

All incoming staff and volunteers should be screened to ensure they do appropriate work. Many workers assigned in Oklahoma City had previous unresolved trauma which was exacerbated by their job experience. Emotional well-being should be frequently assessed with counselling offered if necessary. It is not helpful for long-term relationships to develop between temporary providers and victims. Support should also be made available to the families of disaster workers to better enable them to ease the worker's re-entry into their normal life.

'Project Heartland'

A community's emotional well-being is ultimately the responsibility of local providers. The role of any temporary disaster mental-health effort is to support local providers during the initial phases of disaster when their own ability to respond may be hampered due to the size and type of disaster and the fact that they may be victims themselves.

Planning for the resumption of normal services should begin immediately. In Oklahoma City, within 24 hours of the blast, the Red Cross initiated daily meetings with the State of Oklahoma Department of Mental Health. During these meetings, needs were identified and plans developed for a smooth transition from the Red Cross to "Project Heartland", a special programme set up by the state to provide both short- and long-term counselling programmes for those emotionally affected by this disaster.

Due to the complexity of needs and lack of adequate qualified staff, services were not readily available until six weeks after the explosion. The Red Cross filled this gap in cooperation with the Oklahoma Veterans Administration Hospital and many private therapists, who continued to offer their services on a voluntary basis. The Red Cross provided training for many of these volunteers, using the Red Cross disaster mental-health services course. This assisted them in adapting their mental-health skills to better meet disaster needs.

Many of those who had lost a family member or needed hospital treatment in Oklahoma City were too overwhelmed to seek available Red Cross disaster assistance. The assistance offered included payment of funeral expenses, hospital bills or the costs of bringing family members from outside the area. Outreach teams visited those requesting services in their homes or hospital rooms. As financial needs were met, assessments of other needs were possible. Immediate services and appropriate referrals geared to their needs could then be offered.

Screening and supervision

As well as physical and financial support, the victims of terrorist-caused disasters need emotional support, and disaster recovery is not complete until the emotional wounds have been healed. Disaster workers and organisations will be able to meet the needs of survivors more effectively if they are prepared to deal with their own emotional reactions to the disaster and recovery efforts including the risks of overwork and ownership. Organisations and governmental entities need to learn how to work together to prepare for such incidents. This includes planning the use of volunteers, their screening and supervision.

The emotional toll from the Oklahoma City bombing will be felt for many years to come, as individuals go through the stages of grieving to acceptance. Some may never recover from their loss. Healing the community will also

Sources, references, further information

Devlin, Ed. *Disaster Recovery Journal*, Vol. 8, Issue 3. St. Louis, Missouri, 1995.

Hartsough, D.M., and Myers, D.G. *Disaster Work and Mental Health: Prevention and Control of Stress Among Workers.* National Institute of Mental Health, Rockville, Maryland, 1985.

Hodgekinson, Peter E. and Shapard, Melanie A. "The Impact of Disaster Support Work", *Journal of Traumatic Stress,* Vol. 7, no. 4. New York: Plenum Press, 1994.

Myers, Diane. *Disaster Response and Recovery: A Handbook for Mental Health Professionals.* Rockville, Maryland: US Dept. of Health and Human Services, 1994.

Special Issue on Disasters and Crises: A Mental Health Counseling Perspective. *Journal of Mental Health Counseling,* Vol. 17, no. 3, July 1995.

Weaver, John D. *Disasters: Mental Health Interventions.* Sarasota, Florida: Professional Resource Press, 1995.

Vini Smed, Director, Reference Centre for Psychological Support, Danish Red Cross, P.O. Box 2600, DK-2100 Copenhagen Ö, Denmark. Tel.: (45)(31) 38 1444, fax: (45)(31) 42 1186.

American Red Cross: see Chapter 14.

take time, as divisions persist between those who want to move on and those who will continue to work through their reactions to the disaster.

The American Red Cross and many other agencies, businesses, governmental entities and individuals worked together to meet the overall needs in Oklahoma City and across the country, as everyone involved grieved for the loss of friends, relatives, their community and feelings of security. Knowledge gained in this response must be shared to better enable others to meet the psychological needs of disaster victims and workers in the future. ∎

Railcar refugees: The war in the former Yugoslavia has pushed people first one way, then the other, then back again. Civilians fleeing the fighting may be refugees, displaced people or just plain war victims. Disasters seek out the vulnerable and destroy their lives or livelihoods, so it is those least able to cope – old, young, poor, sick – who suffer most. In a country where family support was a crucial coping mechanism, the elderly – such as this couple living in a converted train carriage – have deep needs.

Old alone, Croatia, 1995. Sebastiao Salgado/Magnum

Bringing relief to those who must flee

Grass grows over the village near the Glina river where old Ivan Mladjenovic once farmed. For four long years in a refugee camp he thought of little but going home. When he got there last summer it had gone. Ivan, 68, can only presume his house was burned. Where its bricks went he cannot imagine. "There is nothing left," he says.

After hostilities broke out in the former Yugoslavia, the fury that swept the Banija region of Croatia in 1991 left it inside the Krajina, part of the one-third of Croatia held by Serbs. Ivan, a Croat, and Marica, his wife, fled with many others. When the Krajina fell to Croatian forces in 1995, he returned to find the village a ruin. Today he sits stone-faced in a refugee camp: "We are too old. How can we rebuild a lifetime's work?"

Ivan's story is not uncommon. Much is destroyed in the area, much will remain beyond the reach of those who fled. Since the fall of the Krajina in August, the changing face of Croatia and Bosnia-Herzegovina has brought further uncertainty to refugees and the displaced.

"The mood is darker now than it has ever been," says Irena Barisic, a Croatian Red Cross social worker in Ivan's camp in Sisak, before August a front-line town. "When people first came here they were shocked and stressed. Then they started to live, started to hope that one day it would all be over. Now maybe it is, but many have no home to go to."

This chapter will look at the impact of displacement and destruction in the Balkans. In the wake of the US-brokered peace agreement made in Dayton, Ohio, in November 1995, all sides in the conflict, and the humanitarian agencies, are trying to cope with millions in need while also embarking on political, economic and social reconstruction.

Even if a peace can hold, and with it a right of return for everyone, the suffering of the Bosnian people – Serb, Croat and Muslim – will continue for many years. The conflict displaced or made refugees more than three million people; hundreds of thousands can never go home, many thousands will not want to begin again.

Most believe the partition of Bosnia-Herzegovina is a fact. The Dayton agreement formally acknowledged a Serb republic and a Muslim-Croat federation. A shell of central government and the device of a federal Bosnian structure cannot hide a country rent apart. As political commentators bicker, the main humanitarian question is clear: What is to happen to all those homeless people?

Some will not return. Will Muslims return to the Zvornik region of eastern Bosnia, where their houses were burned and their mosques destroyed? Can a Croat pick up his life in a village near Banja Luka where his priest was burned alive or his church blown up? Can Serbs return to their Krajina villages where civilians have disappeared?

Dayton theoretically guarantees not only the right of return but the restoration of property, or compensation for it. Those who signed the treaty pledged all could return "in safety, without risk of harassment, intimidation, persecution, or discrimination, particularly on account of their ethnic origin, religious belief, or political opinion". The reality will emerge over time.

Every winter has brought suffering since the conflict began in the former Yugoslavia but the International Federation of Red Cross and Red Crescent Societies and the International Committee of the Red Cross (ICRC) were prepared for the worst yet as 1996 approached. Never had there been a greater refugee crisis and never had the population's resources been more limited.

Scramble for territory

Even excluding Bosnia, the former Yugoslavia started 1995 with 650,000 refugees and displaced people. Summer and autumn saw massive population movements. Ahead of the autumn cease-fire, armies scrambled for territory with which to negotiate. More than 400,000 new victims of conflict had to find somewhere to live.

In the Krajina, deserted by 200,000 Serbs as a Croat offensive began, 9,000 mainly elderly Serbs remained in remote villages. Ethnic minorities expelled from north-western Bosnia were being resettled in the same area. They, too, were destitute.

In central Bosnia, tens of thousands of Croats and Muslims arrived after their expulsion from the fallen United Nations (UN) safe havens of Srebrenica and Zepa, and from the Banja Luka region in the autumn. In Croatia, 21,000 sought sanctuary.

In the Federal Republic of Yugoslavia, authorities were struggling to assimilate 175,000 refugees from Western Slavonia and the Krajina. Under UN sanctions and with limited international aid, its crippled economy was already hosting 430,000 registered refugees.

Around Banja Luka in Serb-held Bosnia, the number of displaced exceeded 150,000 by the end of October. Local authorities had used public buildings, schools, sports halls, private homes, even partially destroyed houses in an attempt to shelter newcomers.

Elsewhere in Serb-held Bosnia were 50,000 highly vulnerable people whose homes had been damaged in the conflict, in some cases by NATO air strikes.

The long-term needs are obvious. Massive reconstruction aid will pour into Bosnia itself, but to rebuild a country requires more than money, bricks and mortar. Society itself needs reconstruction. As the *World Disasters Report* has emphasised before, wheat flour cannot solve political problems in the former Yugoslavia or any other disaster. What is also clear is that political expediency cannot conceal humanitarian needs.

Some problems have been covered by indifference. The suffering of Serb civilians has mostly gone unnoticed. Those leaving the Krajina and Western Slavonia were momentarily news. How displaced Serbs fared later was of little media interest, although the Bosnian Serb authorities did not help. Journalists could not move freely around Bosnian Serb-held territories.

Was there something else? In the *Los Angeles Times*, Alexander Cockburn commented: "When Bosnian Muslims are shelled, driven from their homes or murdered, the world weeps. When Serbs are driven from their homes or are discovered with their throats cut, eyes stay dry. When Serbs do the cleansing, it's 'genocide'. When Serbs are cleansed, it's either silence, or an exultant cry that they had it coming to them."

The Western world – particularly the media – has clearly accorded less sympathy to victims on the Serbian side of the conflict. Journalists, who flocked to Croatia in 1992 and 1993 to document the suffering and despair of displaced Croats and Bosnian refugees, were reluctant or uninterested to report on more than half a million Serb refugees in the Federal Republic of Yugoslavia. Western donors of aid are numerous in the Muslim and Croat regions of Bosnia and few in Serbian ones.

Those on the move faced new challenges in finding homes as a bitter winter approached. Having fled with few possessions, leaving unharvested crops to rot, some were sent to empty and damaged villages. There they had land but nothing to work it with, and often no roof. Others were placed in host family houses. How soon before the displaced became a burden?

Around the north Bosnian towns of Brod, Derventa and Vukosalvje, thousands were living in appalling conditions, some in communities with neither water nor sanitation. In Brod itself, 40 per cent of the town was destroyed and the public water system had a 60 per cent leakage factor. The population doubled with 8,000 new arrivals. Even the water was risky. From underground supplies fed by the Sava river, its very high iron content explained the town's diarrhoea epidemic.

Box 9.1 Bosnia – will there be a brain drain?

When the American college scholarship came through, the raven-haired young woman from Sarajevo agonised over what subject she should choose. Modern English literature appealed greatly but she considered that selfish. "What," she asked a friend from the International Federation, "will Bosnia need most when I have finished studying?" She chose psychology.

As she languished in a refugee camp in Croatia for more than a year, this articulate, intelligent woman saw war trauma, its anguish, pain and devastation, at close quarters. She would make the most of her chance to study again.

From its contacts in more than 280 refugee camps and centres across the country, the Federation was asked to identify capable young people who might want to apply for a small number of four-year American college scholarships. The young Sarajevan, a volunteer interpreter in her camp, was first on the list. Over the past two and a half years she has surpassed herself in the United States; she will probably graduate a year ahead of schedule.

Her proud college has now offered their straight-A's scholar post-graduate studies and sponsorship. She faces a dilemma: should she return to Bosnia or try to stay in the US where she would face a very bright future?

Similar questions are being asked by talented young Bosnians on all sides of the political divide.

Tired of conflict and Balkan politics, opposed to homogeneous society, afraid peace will not last, many abroad will stay there. Many of their friends at home are anxious to emigrate at the first opportunity.

Said a young Bosnian Serb journalist, awaiting NATO air strikes in a Banja Luka shelter last September, "What kind of future is there here for anyone? I have a child. As his mother, I must do my best for him. I don't want him growing up with all this."

Aida, 24, a graphic designer, endured the worst of the siege of Sarajevo. When her elder sister – stranded in Croatia when the conflict began – became ill, she escaped to Zagreb to care for her by running the night-time sniper gauntlet across the airport. When her sister was strong enough, they returned the same way, Aida refusing to desert her home in its hour of need.

As peace came, however, she and her front-line soldier husband left the city they love for the United States. She was six months pregnant. A friend in the US had sponsored them, and her husband had a job waiting for him as a ski instructor. They will work hard and prosper. She was tearful on leaving, telling friends it broke her heart but there was nothing left for them in Bosnia. Sarajevo was no longer Sarajevo; war had rent it apart. Bosnia is the poorer for her departure. ∎

While Bosnian Serb refugees poured in, minority Croats and Muslims were violently forced out to make space in the final chapter of "ethnic cleansing". Where dispossessed confronted dispossessed on the refugee crossroads of the former Yugoslavia, forced displacement spiralled out of control. There were reports of murder; intimidation and beatings were commonplace.

For some elderly Serbs, it was too much. Sudden deaths among old people in collective centres often offered no explanation except physical and psychological exhaustion. They had lost everything, been uprooted two, three, even four times; it was as though they had given up and decided to die.

Escape route

Of the 500,000 Muslims and Croats in north-western Bosnia and the Banja Luka region before the conflict, perhaps 30,000 were left by the end of the summer. Serb arrivals so increased the pace of expulsion that the escape route could not cope. Across the Sava river, the Bosnia-Croatia border, Croat authorities were slowing traffic until those ahead could be found accommodation. Refugee camps were bulging, they said. There was a virtual hold on Muslim entries.

In northern Bosnian towns like Prijedor and Sanski Most, minorities were in crisis. The fear was palpable. With nowhere to go, Muslims expelled from their houses moved in with friends and family who themselves would soon be pushed out.

Agencies in northern Bosnia witnessed this with disbelief. How could those expelled be denied the right to cross the nearest border, rights enshrined in international legislation? It was easy to accuse Croatia, but Croatia had been flooded with refugees and the displaced since the start of the conflict. If other borders opened, refugees could transit through Croatia. The UN High Commissioner for Refugees (UNHCR) reported that it took three months to arrange passage for Muslims to third countries.

Nowhere was the hold-up more obvious than Banja Luka. People huddled under plastic shelters as rain fell, strained faces staring out at the yard of a driving school that had become a makeshift transit centre. Men, women and children, weary and afraid, waited for passage to Croatia.

The transit centre slowly began to resemble a refugee camp. Scared people kept arriving. Police protection made the centre safe, agencies laid on water and sanitation, there was a meal available every day. It was no place for more than an overnight stay; some people stayed for weeks.

The centre straddled the human crossroads. Before people left they would browse through property advertisements pinned to a wall. Krajina Serb refugees placed them there, hoping to exchange legally homes they had fled in Croatia for those Muslims and Croats were fleeing here. It was a risk for both sides. No one knew what had been left standing, or who you would find in the place when you got there. Business was not brisk in Banja Luka. Besides, the authorities would determine many people's destination.

Resettling areas abandoned by the Serbs was a high priority last autumn, in Croatia itself, and in Bosnia. Moving people in behind advancing front lines sometimes assumed grotesque proportions. People's welfare seemed of little account, their security disregarded.

Take the 250 displaced people transported to the Bosnian village of Kondusa, 16 kilometres north-east of Petrovac in September 1995. They found the village still mined, corpses in the houses, and animal carcasses strewn around. Housing elsewhere was so scarce, the authorities argued; they had no choice. This was not easily accepted; the displaced refused to stay.

More subtle social engineering occurred in Croatia. Catholic Church sources there reported that priests expelled from Bosnia were put under pressure to relocate to Croat-held territory with their communities. Agency workers witnessed how two Catholic priests refused such a request from Croatian officials. An ultimatum followed: comply or stay in Bosnia.

In response to all this, the International Federation, ICRC, UNHCR and other agencies expressed concern, and protested vehemently when necessary. All parties to the conflict in the former Yugoslavia have widely violated people's rights to safety and dignity. Having escaped the nightmare of Bosnia, Muslims and Croats were deprived of them again.

One of the strongest protests came from the Federation when, in late September, Croatia's Office for Displaced Persons and Refugees announced it would revoke the refugee status of 100,000 people from Bosnia-Herzegovina as part of a plan to send them home. It was clear that security for returnees could not be guaranteed in many destination areas. The Federation argued that with winter coming the timing of the move was doubly bad, and postponed the launch of a revised appeal for Croatia. After other protests, the Croats eventually compromised on the timing. No refugee would be forced to repatriate, they promised.

The Krajina was being resettled. When Croatian forces launched Operation Storm to regain it from rebel Croatian Serbs in August, most of the population left. A few thousand elderly Serbs remained, to face destitution, intimidation, persecution and even murder. The Krajina was lawless. Homes were burned, livestock roamed freely, and Croat looters emptied Serb houses. Most people were strangers. Even the police were bemused outsiders. As for the new arrivals, they were Croats but most had Bosnian accents.

Box 9.2 Former Yugoslavia – the mines menace

There are six million mines in the former Yugoslavia, three million of them in Croatia. These and unexploded ordnance, says the United Nations (UN), could be the greatest obstacle to a resumption of normal life, from the human impact of injuries and deaths to the costs of mine clearance and the economic impact of unfarmed fields.

The return of civilians to former front-line areas, and arrival of ethnically cleansed settlers in ethnically evacuated regions, heightens the risks. Those who left took any knowledge of mines with them. Those who arrive – including aid workers – pay the price, unless they are very careful.

By mid-1995, the World Health Organization estimated that the conflict in the former Yugoslavia had caused at least 5,000 mine-related amputations. It said the figure excluded deaths and fragmentation injuries, while statistics for Serb-controlled areas were unavailable.

Handicap International reported 20 to 30 amputations a month around Mostar, Tuzla and Zenica. The ICRC noted a disturbing increase; at the start of 1995, four a month were being reported from Zenica but in July the figure was 238. In the former UN sector north – the northern Krajina – monthly numbers rose from 36 in January to 106 in June. Southern section statistics rose from five in January to 20 in May. Casualty figures could soar as people resettle.

The Croatian authorities have moved fast but the UN's Mine Action Centre in Zagreb warned that three million mines across more than 1,000 square kilometres would have long-term and expensive implications. Mine-clearance efforts in Cambodia cost 7 million US dollars a year to clear ten square kilometres. Better infrastructure in Croatia might cut costs, but they will be insignificant.

The problems of land-mines in the Balkans are part of a worldwide scourge where over 110 million land-mines are scattered across 64 countries. They kill or maim over 2,000 people every month. Appalled by this needless suffering, the International Red Cross and Red Crescent Movement has launched a worldwide campaign to get land-mines banned. The slogan of the campaign is direct and simple, "land-mines must be stopped". For more information on the campaign, contact either the International Federation or the ICRC; their addresses can be found in Chapter 14. ∎

Manda Ivascanin, 51, and her three children were living in the small, war-damaged town of Udbina. They had fled Presnace, a village near Banja Luka. St. Theresa's, its Catholic church, had been blown up. In its burned rectory the charred bodies of the village priest, 52-year-old Filip Lukenda, and a 43-year-old nun known as Sister Cecilija, had been found. Some reports claimed they had been doused with petrol and set on fire. Manda says it was the turning point. Villagers got the message: no point in staying.

Tracing request

When the opportunity presented itself in August, she and the children crossed the Sava. Her 53-year-old husband was not with them. Two weeks before, he had been detained for what in the former Yugoslavia was called "working obligation". Mostly that has meant working on front lines. Minority men on all sides have been pressed into digging trenches, clearing mines and labouring in life-threatening situations. International law forbids it but international law has not influenced much in places like Presnace. Manda filed a tracing request with the ICRC; months later there was still no news. He is one of tens of thousands of "missing" men.

Her first concern was survival: finding work, money and winter clothes. Like all refugees, the family had received a small grant from the Croatian government. Free bread was available, the ICRC was distributing food parcels, and the Croatian Red Cross and the Federation were planning a relief programme.

Asked what she hoped for, Manda said, "My husband...jobs for my children." Would she consider going back to Presnace? The answer was long and hesitant, but amounted to, yes, tomorrow, if they could be safe. Presnace, after all, was home, whatever had happened there.

Udbina at least had water and electricity. If the Croatian government were to go ahead with plans to resettle tens of thousands of people in a short period, without seeing to the infrastructure first, another crisis would be in the offing. A Danish Red Cross survey showed the water distribution system to be crumbling. Engineer Bjorn Sorenson said: "The only reason we are not facing an emergency now is that the population is barely trickling in at the moment."

The Dane cited a litany of woes, including a reported 30 to 40 per cent leakage from the public water supply in the southerly Knin area. "Normally in Europe you can accept up to 15 per cent leakage without feeling alarmed," he said. "If you have 60 per cent you had better dig the whole thing up and replace it. Unless something is done here we could be up to 60 per cent very soon."

There was more. The pipe system for the Vrhovine area had been blocked and unused for almost five years, almost certainly rendering it beyond repair. Pipelines in the towns of Licki Osik, Medak and Donji Lapac were badly damaged, and the main supply line for the Benkovac and Zadar regions was said to be losing the critical 60 per cent, most of it from a corroded pipe crossing the Karisnica river.

War damage, deliberate destruction and poor, if any, maintenance has left the southern Krajina in deep trouble. The Dane advised his National Society to implement a maintenance and leak-detection programme fast. The Croatian government had said it wanted to see 80,000 people move into the Krajina within months.

The scale and barbarity of the conflict in the former Yugoslavia should have provided many lessons for the international community. Humanitarian agencies should be wiser, faster and better coordinated. It has been a testing ground for all parts of the International Red Cross and Red Crescent Movement working together, and the Krajina, perhaps, was the ultimate challenge.

Not since World War II had Europe seen such an ethnic exodus. When the Croat offensive began, 200,000 Croatian Serbs living there fled on horse-drawn carts, tractors and trailers, in trucks, cars and buses.

As the first columns of grim-faced refugees flowed into Bosnia, and through Croatia, along the Zagreb-to-Belgrade motorway still labelled the Brotherhood and Unity Highway, the numbers spelled disaster. Shocked and distressed, unprepared for the journey, they would soon be hungry. They had little water and, under a searing sun, they needed it badly. For the very young, the old and the infirm, the ordeal was unbearable. Without shelter, some would not survive.

On the road, Pera Olujic was asking about the husband she had not seen since the offensive began. A refugee from the south-western Krajina town of Obrovac, heading north-east with her daughters in a column of several thousand, she gave birth to a son on the road to Martin Brod.

Box 9.3 Watching for welfare if peace persists

Janja fingers her rosary and mumbles. "When did I leave Gradacac? I forget. The paprikas were turning red." She is 89 – or is it 90? – a confused old lady in the geriatric sick bay of a Slovenian refugee camp. Mostly she sits on her bed with her rosary and black head scarf, mouthing prayers in silence, crying over things she forgets.

Zalkilda, 45, is mentally ill. Her home in Tuzla's Batva district is a filthy hovel without sanitation, and the smell is sickening. When she last washed is impossible to guess. She stares from under the bed sheets, and ignores questions. The curtains are drawn in mid-afternoon. Alone, she seeks darkness. Neighbours sometimes leave food, but the canton's social welfare has ground to a stand-still; Tuzla's Centre for Social Work has no funds.

Camps for refugees and the displaced are full of Janjas, and villages hide many Zalkildas. Old and sick, incontinent and senile, mentally ill and handicapped – all abandoned or lost by families. No one wanted them, least of all so-called third countries offering refugee places.

Who will care for these people if peace persists? Care for the elderly with trained staff should be a priority. Given the collapse of social welfare, it is unrealistic to expect the republics of former Yugoslavia to cope alone. As others rebuild, Janja, Zalkilda and the rest face a new crisis.

Zalkilda talks to one person, Sala Nakic, a Tuzla Red Cross volunteer. Sala says Zalkilda has been ill a long time, and has developed respiratory problems: "If there hadn't been a war she'd be in an institution. The thing is, right now, the place is across the former confrontation line."

Sala is part of a Swiss Red Cross social-care programme implemented by Tuzla's Red Cross branches. It offers food assistance – 5 kg parcels of ready-to-eat meals, dehydrated soups, stew, salt, cheese – and gives volunteers access to the homes of the most vulnerable.

Across the Tuzla canton, the programme identified 7,000 people without help from family, state or relief agency. Old, alone, paralysed or handicapped, some are mentally ill, none can look after themselves.

The Swiss Red Cross mobilised all 14 municipal Red Cross branches, and up to 500 volunteers make visits every fortnight, giving each beneficiary a food parcel and what Michel Paris, the Swiss Red Cross coordinator, describes as a regular social presence.

Across the former Yugoslavia, delivering relief brings contact with the vulnerable and an opportunity to provide more protection. In Croatia, after most Croatian Serbs fled the Krajina, 9,000 mostly elderly Serbs remained behind, too weak or too obstinate to run. All were vulnerable, given a hostile human environment, the lack of community support, and the approaching winter.

Mobile Red Cross teams went out to find these people. They distribute food and hygiene supplies, provide medical care and protection, and report harassment to the Croat authorities. The contact is vital. Who else would visit 84-year-old Marija Radinovic? Someone burned and looted houses in her deserted Radinovici hamlet but Marija would not dream of leaving.

Her house is a low, one-roomed, stone-floored dwelling. Up in the loft she has a little food, dried berries for tea, and some clothing. In the yard a few chickens scratch. She grows vegetables, there is water; it is enough, she says. It isn't, but a Croatian Red Cross relief team will keep an eye on her. Relief can have added value. ■

Her contractions had begun at 4h00 as she lay in the back of a truck. She had got off to search for medical help, and was found by an ambulance, with a nurse who was tending a wounded soldier and some children who had lost their parents. They squeezed up to make room. The delivery took four hours, and even then her ordeal was not over. There was no water to wash the baby.

Hours later in the west Bosnian town of Petrovac she would find relief. People with red crosses on their aprons gave her food and water, and transport to a Banja Luka hospital bed.

Even before the refugees crossed into Bosnia, the Movement was mobilising. When the ICRC in the Krajina capital of Knin reported the first shelling, the need for a rapid relief response was confirmed. The ICRC itself, with its mandate for action inside conflict areas, and the local Red Cross in Bosnian Serb-held territory, responded to the immediate crisis. The Croatian Red Cross, Yugoslav Red Cross and the Federation prepared for its rapid spread.

The ICRC strategy of decentralised relief stocks bought time. Enough food parcels, bulk food, survival rations, blankets, tents, candles, soap and hygiene items were on the spot throughout the region to cover a major emergency for about ten days. Red Cross volunteers poured in to help. Unforeseen, however, was the access problem, with roads blocked by a mass of humanity.

Water was a major problem. The ICRC used an 18,000-litre tanker to replenish water points along the roads from Petrovac in the west to Bijeljina in the east. Journeys were very slow but the water kept coming 24 hours a day from ICRC's hub in Banja Luka. Perhaps 30,000 refugees stayed in the Banja Luka region, while the rest continued eastwards. By the end of August, 175,000 people had crossed the border of the Federal Republic of Yugoslavia.

For the Federation and the Yugoslav Red Cross, it was the start of a long campaign. While resources were released from other programmes, the Federation appealed to the international community for more than 5 million Swiss francs (Sfr.).

Reception points

The entire staff of the Yugoslav Red Cross and 8,500 volunteers provided 24-hour assistance for new arrivals. At five border crossings opened for the influx, reception points provided food, water, hygienic goods and first aid. All main roads into the country had support points. Relief was then channelled to transit centres, often in schools or sports halls, and 217 Yugoslav Red Cross branches cared for those moving on to host families and to collective centres.

Where host families and centres could not cope, Yugoslav Red Cross soup kitchens throughout the republic responded. The extent of operations in Croatia, meanwhile, was reflected in another appeal from the Federation. With the Croatian Red Cross, it asked for Sfr. 2.5 million in September. In October, the figure was revised upwards to Sfr. 5.6 million.

The Croatian Red Cross was under pressure from all directions. While it assisted Serbs fleeing the Krajina through Croatian territory, it faced the fallout of the Serb influx into northern Bosnia. It had to ease the lot of Croats and Muslims being expelled over the Sava as a consequence. Having provided fuel for the ferry boats plying the escape route, it set up feeding stations for the refugees. It ended up caring for 21,000 Bosnian Croats and Muslims.

On the road from the Krajina, some looked as if they would not make it. There were people who had walked most of the way, others camped by the road, some stuck in broken-down vehicles. Early on 15 August, a Red Cross convoy of 23 buses left Belgrade for Banja Luka to find them. Forty-eight hours later it was back, with 1,420 people on board – the elderly, wounded and sick, women, babies and small children. Next morning the buses would go out again.

Sources, references, further information

Glenny, Misha. *The rebirth of history: Eastern Europe in the age of democracy*. London: Penguin Books, 1993.

Glenny, Misha. *The fall of Yugoslavia: The third Balkan War*. London: Penguin Books, 1994.

Milivojevic, Marko. "Bosnia-Hercegovina; Serbia, Montenegro", *World of Information Europe Review 1996*. London: Kogan Page, 1995.

Minear, L. et al. *Humanitarian Action in the Former Yugoslavia: The UN's Role 1991-1993*. Thomas J. Watson Jr Institute for International Studies, Providence, USA, 1994.

Former Yugoslavia on WWW: http://www.wideopen.igc.org//balkans/webl.html

Red Cross and Red Crescent Societies in the former Yugoslavia: see Chapter 14.

International Federation delegations in the former Yugoslavia: see Chapter 15.

The swift and determined response of the Yugoslav Red Cross eventually meant that over 80 per cent of all international humanitarian aid to the refugees was channelled through the network it had built up over recent years with the Federation.

For all that, the outlook was disturbing. Many refugees were elderly, and many were children. A bitterly cold winter was approaching, and clothes, shoes, beds and stoves were badly needed. All but 20 per cent of the refugees were placed with host families, the rest went to collective centres, but many host families were themselves unable to cope; requests for transfers to collectives grew.

Grim situation

Dr Rade Dubajic, Secretary General of the Yugoslav Red Cross, put it bluntly: "We have a difficult economic situation. When you add to that a total of 600,000 refugees and around two million other people surviving in below-minimum living conditions, we face a grim social situation. No more collective centres are available, and those that exist are inadequately equipped. Yugoslavia has no funds to improve them, or to build new ones."

On 8 September, the Federation revised its appeal: Sfr. 19.7 million was now needed to see 337,000 people through the winter. It had been a long, hard road from the Krajina, but the journey was not over yet.

The Krajina crisis highlighted the barbarity of the conflict, and the blatant disregard for the traditional distinction between combatant and non-combatant, cornerstone of international humanitarian law. Retribution in the Krajina, the ill-treatment of the elderly Serbs who declined to run, was one aspect. The targeting of refugees in flight was another.

On the day of the offensive, Obren Vukovic and his two grandchildren scrambled on to a bus in a refugee column leaving the little town of Donji Lapac. During a stop, Obren's dark mood brightened when he saw his son drive up in a truck. Obren climbed on board with his grandchildren.

A little further on, the column was attacked. Obren does not remember what happened as he lost consciousness. He came round to find himself on the roadside and his son's truck burning furiously. He lunged towards it to pull his family out. People held him back; it was too late.

Obren was taken to hospital in Petrovac to have his own wounds treated, but the following day he returned to the scene of the attack. He found the truck and collected in a small plastic bag all that remained of his family. Then he dug a little grave and buried it.

Obren is now in Banja Luka, physically recovered, as one day the former Yugoslavia and its peoples may be. But, he told his doctor, his most serious wound was to his heart, and that one would never heal. ■

Clear signals: Good disaster response is dependent upon good information. For many international humanitarian agencies, working for the first time in DPR Korea has been a steep learning curve. Amid an emergency of devastating floods affecting millions and against a backdrop of growing hunger, information on the country, its culture and basic structures was difficult to gather. Local information on the needs of disaster survivors was often mistrusted by outside organisations and donors.

Controlling the traffic, DPR Korea, 1988. Hiroji Kubota/Magnum

Partnership and politics in disaster

E ven in a year that saw floods in most Asian nations, the floods that hit the Democratic People's Republic of Korea (DPR Korea) in 1995 made headlines around the world. The severity of the disaster seemed appalling, and – news itself – the government of a traditionally very secretive country provided detailed information, assessed that it needed extra resources to cope with the situation and appealed to the international community for help.

In a few days in August, 800 millimetres of rain poured on 145 of the country's 200 counties. In some regions near the Chinese border the rainfall exceeded 100 millimetres in an hour, one-tenth of the average yearly rainfall in the country.

In mountainous areas the rain turned into a roaring wave that swept away houses, roads and bridges. In Sinuiju, a large town on the northern border with China, the river Amnok broke its banks, flooding thousands of hectares of paddy fields with waves of mud that then dried, turning fields into desert. In Rinsan county, 80 kilometres south of the capital Pyongyang, 12 million cubic metres of water flooded villages as the dam of Sangwol reservoir collapsed.

On government estimates, the disaster affected more than five million people, including half a million left homeless, and damaged 40 per cent of arable land. Hundreds of thousands of people were injured, but no official figures were given for the casualties, and different sources quoted death figures of between 60 and nearly 300. Given the extent of the physical devastation and experience from other major floods, the higher death toll estimates are more likely to be correct.

This chapter will examine the impact of the flood disaster in the DPR Korea, the response operation by national and international agencies, and the broader context of the country's economy, agriculture and foreign relations, especially its so-called "disaster-aid diplomacy".

Survivors of many sorts of natural-trigger disasters face acute food shortages. As a standard procedure, relief agencies often include a substantial food component in their plans, aiming to supplement the diet of survivors for two or three months. In early December 1995, it became clear that for the DPR Korea floods this was not sufficient. Far more emergency food aid would be needed – around 1.16 million tonnes of grain, one-tenth of the entire world's food-aid supply – to feed the DPR Korea for much longer.

To understand why so much food aid is needed for so long, one must look at how food is managed in a centrally planned economy and how severely the floods affected the communal distribution system.

In the DPR Korea, the government defined "norms" for basic consumption needs, calculated in grams per day for various groups, with 75 per cent of calories provided by cereals, almost exclusively rice. In urban areas, people receive their rations from public distribution centres twice a month.

Those in rural areas – who were more than three-quarters of flood survivors – are in a different position. Each family retains its quota when handing over the crop, harvested between October and December, to communal farms. In January 1995, farming families handed in the 1994 harvest, retaining their ration for the next 12 months. When floods hit in August, it is estimated that an average family of five had around 200 kg of rice in stock. These stocks were washed away together with the houses and all belongings.

Families without food could not replenish their stocks, since the same floods destroyed communal silos and warehouses. The floods created the conditions for a major nutritional emergency in a third way by spoiling nearly 40 per cent of arable land within months of the 1995 harvest. Nearly one million tonnes of rice were missing from the expected 1995 crop.

Facing starvation

The prospect was frightening; a substantial part of the farming population was facing long months of starvation. Food aid was badly needed to take this population through to the 1996 harvest. When agencies started reacting to these needs, the government showed unexpected tolerance *vis-à-vis* the extensive media coverage – not always friendly – that the emergency attracted.

To the surprise of many, no consequences were reported after a senior official from an aid agency in Pyongyang made politically insensitive remarks to the BBC World Service. The government itself kept speaking openly about the severity of the situation and repeating its readiness to accept help.

In early 1996, the situation was still developing. At the new year's reception, the vice-minister of foreign affairs repeated his dramatic appeal: "I visited the Pyongyang thermal power station this morning. Out of 13 generators, only three were working. We do not have enough coal because the coal mines were filled with water following the floods. Nevertheless, we think that we can deal with these problems. We need your help with food." That summarised the official government position: floods caused all the problems and the problems were serious.

As news spread in late 1995 of a drastic reduction in the food rations of the general population, the government's analysis of the causes of the problems was placed in a broader context by a report of a joint WFP/FAO food-security survey. From direct observation and with data supplied by various ministries, the United Nations (UN) experts showed that the food shortage was only partly due to floods; the deficit incorporated a structural component.

Maize output

According to the report, between 1989 and 1993 rice production declined an average of 3 per cent annually, while maize output fell 6 per cent annually. The agricultural decline followed strictly the decline of a collapsing economy. It was estimated that rice production fell another 10 per cent between 1993 and 1995, 15 per cent for maize. Per capita availability of food grains from domestic production was 345 kg in 1989, 272 kg in 1993, and only 222 kg in 1994, the year before the floods.

What could be expected from publication of this information, implicitly demonstrating that the agricultural policy was wrong and recommending a sharp change? Many expected angry reactions. On the contrary, the government welcomed a further FAO mission, whose mandate was to advise on alternatives to monoculture and land exploitation. Some have interpreted this as strong evidence to support the theory of disaster aid being used as a diplomatic tool. Others take a more pragmatic line.

Early in the relief operation, the national media reported widely on the activities of these international assistance agencies, emphasising foreign countries were providing assistance. Small steps but steps nonetheless, given that official statements can ferociously criticise some of those countries.

The attitude of the government in dealing with the agencies has been regarded as an important sign. Aware that good agency reports to their donors is one way to maintain aid, the government has been extremely

Box 10.1 DPR Korea – stories of survival

The temperature is falling fast as night closes in on a jumble of mud-brick huts. They were built by hand after floods swept everything away in this hamlet, which sits in the shadow of the mountains near the Chinese border. Food is tightly rationed yet the hungriest months – August and September, just before the autumn harvest – are still to come in the Democratic People's Republic of Korea (DPR Korea).

"For many, it's a grim and miserable existence," sums up Paek Yong Ho, Secretary-General of the Red Cross Society of the DPR Korea, which swiftly stepped in to help hundreds of thousands of people across the country.

With millions of Swiss francs from other National Societies worldwide, Paek's staff and thousands of Red Cross volunteers have distributed clothes, blankets, medicines, building materials and food.

"All the Red Cross beneficiaries were left with just the clothes on their backs. They suffered a total loss, their homes, possessions, food stocks – everything," adds Paek.

Each flood survivor tells their own story of disaster. Ho Song Hun, 36, a widowed mother of three, remembers the sound of helicopters as the floods rose in her Sinuiju City home.

"At the time, I was at a loss and feared that all of us would die. So all I could do was cry. After the flood, I lost everything, my property and food. I was thinking, 'Will we starve?'

"But we have received some rations from the Red Cross which we mix and boil with vegetables. With this food we can keep our lives going for the time being," says Ho.

Orphan Ki Hyang, 12, is lucky to be alive. A neighbour and Red Cross volunteer, Ki Gyang Hui, snatched Hyang from her bed by as waters sudden-

ly engulfed her family home near Huichon City, killing both her parents.

Now she is a member of her rescuer's family: "I was very sorrowful when I lost my mother and father – my heart was broken. But now my kind neighbours look after me like my real parents – we are a strong family."

Chu Jong Chol, a 47-year-old factory worker now lives with his wife and three young children in a mud-brick house in northern DPR Korea: "After the flood we moved to one room in our factory for a month. The Red Cross provided us with basic necessities and recently we have received blankets, quilts and rice. Our new house is still damp and we feel cold.

"At present there is very little firewood around due to the floods and it's difficult to keep warm."

The International Federation's response system allowed it to channel resources from across the world – 169 National Red Cross and Red Crescent Societies – to the National Society in DPR Korea to assist flood victims.

It also proved the quality of work in DPR Korea. Dr Piero Calvi-Parisetti, the Federation Head of Delegation in Pyongyang, speaks from long experience.

"There is only one word to describe how the Red Cross Society of the DPR Korea is organising this flood-relief operation – outstanding. It is hard to believe it is the first time the Society has had to undertake a post-disaster programme on this scale and intensity.

"The speed and efficiency with which relief goods have been transported from the point of entry to the target areas and then distributed fully in accordance with Federation guidelines has been remarkable," says Calvi-Parisetti. ∎

cooperative; providing information, resources, and, what is most important, putting no restrictions on access to all flood-affected areas and beneficiaries for monitoring.

The WFP, UN Department of Humanitarian Affairs (UNDHA) and the International Federation of Red Cross and Red Crescent Societies were allowed to travel nationwide at any time to check distributions, make further assessments, and talk directly to recipients. Members of the inviting organisation always accompanied visits, but no pressure was exercised on monitors. Agencies could select monitoring sites at random, and then choose, household by household, to whom they wanted to speak.

As a result, agencies sent back two unequivocal messages to their head offices: that 100 per cent of the aid is reaching those really in need, but the situation is far worse than anticipated.

The suggestion that the impact of the floods might have been underestimated in the initial assessments first appeared among aid workers at the end of November 1995. Some reports from affected areas said that the only food seen in the newly built mud-brick houses was the humanitarian aid. Other reports spoke of people digging for roots in fields. One DPR Korea Red Cross beneficiary told monitors: "If the Red Cross aid stops, we don't know how we are going to survive." The words were echoed by embarrassed mayors and civil servants at local level.

Donor pledges

As information arrived, agencies came to understand the problem and the mechanisms that helped create it. Analysis first focused on flood survivors; half a million people left homeless and, in the words of Trevor Page, WFP country director, "on the brink of famine".

Needs assessment and analysis led to assistance appeals being issued, but pleas for help faced a poor response. While the International Federation closed the year with just over 70 per cent of its emergency appeal met, UNDHA and other UN agencies found themselves in a difficult position. In December 1995, WFP announced that it would close its Pyongyang office if no firm donor pledges were made by early 1996. Three months later, enough pledges had come in to ensure that the office was not forced to close, but the food-aid situation remained of concern.

Since almost all humanitarian donor countries have market economies, the unsatisfactory response has been attributed to the poor profile of the DPR Korea in the West. Yet Ethiopia, under its former Marxist government, received massive US humanitarian aid in the past. More important are the changes in the humanitarian world in the past decade, especially donor demands for high quality monitoring. Building a reputation for allowing good reporting took several months in the DPR Korea.

In late December 1995, some changes started to be seen. Separate public awareness campaigns by the Federation and WFP focused attention on the combined effects of the floods and the underlying structural food deficit. At the same time, donors were receiving information about the first-rate quality of the relief operations and excellent independent reporting. Requests for information and donor pledges increased in early 1996.

Nutrition survey

Some requests were for an independent nutritional survey. This had always been an extremely sensitive issue. Negotiations took weeks, with the aid agencies underlining the need for independent scientific investigations to mobilise funds. Once more, the government made concessions and the survey was planned, coordinated by WFP.

Humanitarian assistance has always operated in a political environment but for the DPR Korea floods, the politics often seemed to dominate the agenda.

For 40 years, since the end of the Korean War, contacts between the DPR Korea and the rest of the world have been very limited. A socialist country with ties to China and the countries of the Soviet sphere of influence, since the mid-1950s its image in the West mostly came from intelligence reports. That lack of contact and information worked both ways; few in the DPR Korea had detailed knowledge of the world outside.

Box 10.2 Meeting food needs today and tomorrow

United Nations (UN) agriculture experts are clear in assessing the problems of the Democratic People's Republic of Korea (DPR Korea) and its needs in the short- and long-term.

They predict a 1996 shortfall of around 1.9 million tonnes of cereals, 700,000 tonnes of which will be met by commercial imports, leaving about 1.16 million tonnes to be found as food aid, the equivalent of one-tenth of the entire global food-aid supply. The UN report clearly warns that the crunch months will be August and September, just before the harvest.

The Food and Agriculture Organization (FAO) and the World Food Programme (WFP) assessment mission report suggests malnutrition will rise through the year unless there is substantial food aid.

Among DPR Korea's total population of 22 million, most at risk are the 2.1 million young children and 450,000 pregnant and nursing women. There is concern for the entire farming population in flood areas; they have lost crops and food stocks yet are outside the urban rationing system. The three provinces most affected and hosting most people in need are Chagang, North Pyongan and North Hwanghae.

Falling food output brought ration system revisions. It once had nine levels, from 900 grams of cereals a day for industrial workers to 100 grams for kindergarten children. Today a three-level system, determined by age, offers an average 2,131 kcals a day, with 60 per cent rice, 40 per cent maize.

Based on that ration, each person needs 100 kg of rice and 67 kg of maize, a total of 3.69 million tonnes of grain in 1996 for human consumption. Even if the emergency forces cuts in the levels of grain required for animal feed and industrial use, the mission estimates total grain needs at 5.99 million tonnes, set against a total domestic output of 4.1 million tonnes.

It suggests that the cause of present shortfalls can be divided almost equally between floods and structural problems.

They set out the causes and consequence of the steep decline in agricultural output from 8.1 million tonnes in 1989 (an average yield per hectare of 6,000 kg) to 6.64 million tonnes in 1993 (4,950 kg) and 4.93 million tonnes in 1995 (3,663 kg).

The substantial structural deficit is caused by both stagnating agriculture and a declining economy. The report says low foreign reserves, large and persistent trade deficits and low credit worthiness have limited both food imports and agricultural inputs.

The break-up of the Soviet Union and other political changes in the 1990s left the DPR Korea economically isolated.

The country has faced other difficulties. Only 20 per cent of this mountainous land can be cultivated, the climate allows only one cropping season and soil fertility has been declining. Two years of poor weather has coincided with rising prices for grain imports, while China, a key supplier, has turned from being a major grain exporter to a large importer.

These problems helped reduce the food-grain stock from four million tonnes in 1990 to nothing today.

But FAO and WFP have grounds for some optimism in looking for long-term solutions. Land areas for cultivation can be expanded by terracing for maize and reclaiming tidal areas for paddy production, which could – at an admittedly high cost – add half a million hectares to the 1.43 million already used for cereals.

The centralised economy and principle of *juche* or self-reliance makes the DPRK almost ideal for food-for-work schemes to repair flood damage and begin expanding the cultivation area.

While the mission doubts that the country could, even in normal conditions, achieve food self-sufficiency, it suggests ways to improve food security and productivity. These include crop rotation and diversification, improved varieties, and the integration of agriculture and livestock farming. ∎

A key factor in – and result of – such isolation has been the philosophy of self-reliance, which in the DPR Korea took the form of the *juche* idea. This theory maintains that "the man is the only master of his destiny" and "all resources must be found within man himself". Applying these concepts to the nation, the DPR Korea recovered from the devastation of the war and went through a remarkable industrial development during the 1960s and 1970s, relying almost entirely on its own resources, but in almost complete seclusion.

The past decade of dramatic changes in geopolitics took their toll on the centrally planned DPR Korea economy. After the dissolution of the former Soviet Union and the change of Chinese policy on economic development, the DPR Korea lost its two biggest – and almost only – trade partners.

A rapidly declining economy, with low foreign exchange reserves, a large and persistent trade deficit and a difficult credit position, has led to a seemingly unstoppable collapse of the gross domestic product since 1989. The DPR Korea now faces serious shortages in raw materials and energy supplies, so most industries are partially or totally idle, electricity for the population is rationed and centralised heating is not available, even in central Pyongyang.

Economic links

Some commentators cynically suggested that the floods came at the right time. In their view, the DPR Korea is at a crossroads. The government has to choose between the policy of self-isolation and an uncertain future, or to allow economic – and political – links with the international community, a choice which elsewhere in the region has proved very successful. Was the request for humanitarian aid the first step towards the latter option?

A partial confirmation of the fact that the government's request for assistance might carry a different weight could be found in the appeal itself. The various assessment teams that visited the country in the immediate aftermath of the disaster expressed some reservations on the governmental estimates of damage.

While there was consensus that 500,000 homeless and 400,000 hectares of arable land damaged seemed quite realistic figures, more than one agency refused to comment on the government estimate of 15 billion US dollars infrastructure damage. This evaluation – equivalent to three-quarters of the nation's gross national product – has been read as a way of "saying without saying", of letting the world know that the country suffers structural problems, which the floods dramatically worsened.

Disaster mandates

If these were at least partially the intentions of the government, then humanitarian agencies working in the DPR Korea – the Federation, the UN assistance agencies, *Médecins sans Frontières*, Caritas, Swiss Disaster Relief and others – would be in a very delicate position. Their various mandates focus on assistance to disaster survivors, not interventions at "macro" level, let alone promoting political and economical links. Speculations on any different role as "ambassadors" or "facilitators" of international relationships at another level leave aid workers uneasy.

There is some evidence, however, that what has been dubbed "disaster-aid diplomacy" might be at work, both in the DPR Korea and abroad. The first indication: the number of foreigners in the country has never been higher.

Relations between the aid agencies and their national counterparts – the DPR Korea Red Cross for the Federation and the governmental Central Flood

Damage Rehabilitation Committee for all others – started somewhat formally. For an entry visa, foreigners must be invited by an "inviting organisation", which retains full responsibility – and control – over the visitor throughout the stay. A "guide" is assigned to the visitor to "facilitate" all aspects of their professional work and many aspects of their free time.

While full access is granted to all parts of the country for business trips, personal movements are de facto strictly limited. Foreigners are not allowed to drive a car, and, although a prohibition has never been made explicit, strolling around downtown Pyongyang is not encouraged. The expatriates, mostly Western, found the situation difficult. Time allowed a better mutual understanding and some degree of relaxation on both sides. A few months on, confidence building – between people and organisations – had progressed, albeit slowly and with the prospects of reversals on both sides.

This process is made easier by the very high level of professionalism and commitment shown by the national counterpart individuals and organisations. When the Federation arrived, very little was known about the structure and operational capacity of the Red Cross Society of the DPR Korea. Delegates did not know what to expect from an organisation claiming a membership of well over one million and more than 300,000 volunteers.

It turned out to be a very strong organisation, deeply rooted in Korean society. An active and unbureaucratic Pyongyang headquarters coordinates

Box 10.3 Viet Nam – unsustainable flood relief?

For most relief organisations 1995 was the first year they worked in DPR Korea, but other countries of the region suffer repeated and severe flooding. Is there anything to learn from relief operations in neighbouring countries?

Viet Nam is one such country. In recent years (1991, 1994 and 1995) the Red Cross of Viet Nam has engaged in three major relief operations, twice with international support, after floods in very low-lying areas of the Mekong Delta affected by high waters in the annual rainy season between August and November.

It found that better-off inhabitants have developed coping mechanisms to prepare for and mitigate the effects of floods. They are able to raise the foundations of their homes, and they are not dependent on a daily income to survive. Landless poor people, however, have little margin to cope. In addition to damage to their homes, the poor suffer most because floods cut their access to all elements crucial to their survival – food, fuel and income – by preventing them fishing, collecting wild vegetables, cutting firewood and working as day labourers.

Rather than providing food, the Red Cross of Viet Nam responded to local requests by focusing relief efforts on income support by distributing boats to the most vulnerable villagers, helping them to fish, collect food, find wood and search for work.

While successful in many respects, there are doubts about the sustainability of these relief operations. The poor in the Mekong Delta have always relied heavily on common resources from land, forests and the rivers. Population growth and more people without land are putting pressure on these resources, while increased pesticide use is affecting fish stocks.

Relief operations are effectively supporting the population to exploit resources which are already being used unsustainably. As this problem increases, the Red Cross of Viet Nam will have to consider supporting alternative survival strategies.

Exploring rehabilitation strategies could lead the Red Cross of Viet Nam farther from its traditional roles.

While the problems are central to its overall goals and mission, the potential solutions (income generation and agricultural activities) may well lie outside its current technical expertise and its locally perceived operating mandate.

Vulnerable people saw subsistence as a priority, more important than health, for example. Subsistence problems rarely have sustainable solutions that can be swiftly packaged and delivered. So should the Red Cross of Viet Nam begin to address the "real" problem if it cannot be handled through its existing distribution and health-service structures?

This issue is at the heart of the debate about whether National Societies and other relief organisations are ready to pursue the much-lauded goals of programming relief for development. ■

the actions of committees spread all over the country at all levels, from regional to city and county, *ri* and *dong*. A *ri* – in rural areas – or *dong* - in cities – is a community of about 300 to 400 families and forms the elementary unit in social organisation of the DPR Korea.

The Red Cross Society of the DPR Korea is involved in most traditional National Society activities, from first aid to nurse training, from youth programmes to disseminating basic hygiene messages. Through a large network of first aid posts, it provides basic health care in remote districts, schools and factories.

Most impressive has been the Red Cross Society's operational capacity. Creating an efficient distribution network, from central warehouses to isolated distribution points, is never easy but the Red Cross Society of the DPR Korea did an outstanding job from the start.

Taking advantage of considerable government logistical support, yet maintaining independence and transparency, the National Society organised distribution of commodities made available by the Federation. Hundreds of thousands of relief items and three and a half thousand tonnes of rice were distributed to 130,000 beneficiaries in areas most severely affected by the floods.

Because of this remarkable operational capacity, the Federation was able to keep expatriate numbers to a minimum, with delegates dealing almost exclusively with monitoring and reporting.

Confidence building – a possible by-product of disaster-aid diplomacy – was fostered by mutual understanding and professional esteem. The confidence is not only being built between colleagues: a broader process is underway.

Aid workers have on occasion been the first foreigners to visit some areas since the Korean War. What did local people do, seeing these strangers for the first time? They smiled, clapped, and offered fruit and small gifts. For decades, the idea of foreigners has been associated with diffidence and suspicion; some nationalities are never mentioned without being preceded by the word "imperialist". Today, foreigners are seen delivering aid, caring for people.

Extremely effective

In humanitarian assistance, as in most fields, those donating resources have to be convinced that a real need exists before international support is offered. Aid agencies and donors have a tendency to believe that which is familiar and generated from within their own ranks. Equally they are suspicious of evidence coming from "interested parties" and unfamiliar sources.

In the mid-1980s the famine early warnings issued by the Ethiopian government's Relief and Rehabilitation Administration (RRA) were played down pending "independent assessments". By the end of the 1980s, the RRA's assessments were proved accurate and timely.

In Iran, following the 1991 earthquake and later refugee-assistance operations, donors were sceptical of the ability of the Red Crescent Society of Iran to deliver assistance efficiently and effectively. The Red Crescent turned out to be extremely effective.

For assistance agencies which are comprised of local organisations and espouse a bottom-up approach to relief work, the donor community's reluctance to trust local sources presents a clear challenge. Whether working in Ethiopia, Iran or DPR Korea, the rationale of working with and through local institutions in order to reach the most vulnerable people in a sustained fashion remains true. Part of the role of the international assistance agency must surely be to champion the use of such local resources and to back up and add credibility to local assessments.

Sources, references, further information

Hoare, James. *North Korea*. Oxford: Oxford University Press, 1995.

McCarter, James. "North Korea", *World of Information Asia & Pacific Review 1996*. London: Kogan Page, 1995.

Red Cross Society of the DPR Korea. *Faces of the Vulnerable. (Spotlight on the National Society and its work with flood victims)*. International Federation of Red Cross and Red Crescent Societies, Geneva, 1996.

Red Cross Society of the DPR Korea: see Chapter 14.

DPR Korea on WWW: http://www.city.net/ countries/north_korea/

Was disaster-aid diplomacy at work between the DPR Korea and the rest of the world – and if so, does it work? No aid worker in Pyongyang wants to speak officially about a subject with such wide implications. Careful observers noted that the change in donor response coincided with other important signals, including the signing of an agreement between the DPR Korea, the Republic of Korea, Japan and the United States on the supply of nuclear reactors to the DPR Korea, and the release of five Republic of Korea fishermen previously detained by the DPR Korea. Amid such intricacies, there is optimism that support will continue for the 500,000 people who lost everything in the floods of August 1995, including a new Federation appeal in March 1996 for more than 7 million Swiss francs to support the DPR Korea programme. ∎

Mass action: Disaster data emphasises the numbers involved. Those numbers are often getting larger, not because there are more natural events – earthquakes, for example – that can lead to disasters, but because population growth and the actions of people put more at risk. Behind the numbers are individual men, women and children, each with a story to tell, a life to lead or a livelihood to lose. Getting the numbers right helps improve the efficiency and effectiveness with which humanitarian agencies serve the people behind the numbers.

Awaiting food, Rwanda, 1994. Sebastiao Salgado/Magnum

Good data for effective response

As disasters become more complex, and the cost of prevention grows, the need for systematic data for disaster evaluation and management has been of increasing concern to international and national relief agencies, governments and donors. Despite the obvious time pressure of relief actions and fundraising, the ad hoc collection of information by individual organisations at the time of an emergency is clearly inadequate for effective disaster response, good management and strategic planning.

In this chapter, the *World Disasters Report 1996* presents data from various sources. The parameters defining the information contained in the tables is given below.

Tables 1 to 11

The Centre for Research on the Epidemiology of Disasters (CRED), at the Department of Public Health, Catholic University of Louvain (Belgium), has developed a system of databases for global disaster management, drawing on its existing disaster documentation, information network and computer system. Tables 1 to 11 of this database section were derived from the EM-DAT, a disaster-events database developed by CRED and sponsored by the International Federation of Red Cross and Red Crescent Societies, World Health Organization, United Nations Department for Humanitarian Affairs (UNDHA), European Community Humanitarian Office and the International Decade for Natural Disaster Reduction.

EM-DAT is now fully operational, with more than 10,500 records of disaster events from 1900 onward, and its own menu for updates, modification and retrieval. Designed to have the right level of detail for wide use, the entries are constantly being reviewed for redundancies, inconsistencies and the completion of missing data.

The criteria for entry of an event is ten deaths, and/or 100 affected, and/or an appeal for assistance. In cases of conflicting information, priority is given to data from governments of affected countries, followed by UNDHA, and then the US Office for Foreign Disaster Assistance. Agreement between any two of these sources takes precedence over the third. This priority is not a reflexion on the quality or value of the data, but the recognition that most reporting sources have vested interests, and figures may be affected by socio-political considerations.

Despite efforts to verify, cross-check and review data, its quality can only be as good as the reporting system. If field agencies were to adhere to standard reporting methods, it would be far easier – and much less costly – to assemble essential data from numerous sources. Data presented in these tables are recorded at CRED; while no responsibility can be taken for a figure, its source can always be provided.

Dates can be a source of ambiguity. The declared date for a famine, for example, is both necessary and meaningless – famines do not occur on a single day. In such cases, the date the appropriate body declares an official emergency has been used.

Figures for those "killed" in disasters should include all confirmed dead, and all missing and presumed dead. Frequently, in the immediate aftermath of a disaster, the number of "missing" is not included, but it may be added later. Without international standards, definitions vary from source to source, so each entry is checked for clarification.

People "injured" covers those with physical injury, trauma or illness requiring medical treatment as a direct result of disaster. First aid and other care by volunteers or medical personnel is often the main form of treatment provided at the disaster site, but it has not been defined whether people receiving these services should be included as "injured".

"Homeless" is defined as the number of people needing immediate assistance with shelter. Discrepancies may arise when source figures refer to either individuals or families. Average family sizes for the disaster region are used to reach consistent figures referring to individuals.

Defining "persons affected" is extremely arduous. Figures will always rely on estimates, as there are many different standards, especially in major famines and in the complex disasters of the former Soviet Union and Eastern Europe.

Disparities in reporting units can create dilemmas, such as the monetary value of damages expressed in either US dollars or local currencies. While it is easier to leave currencies as they are reported and convert them only when the event is of interest, this procedure effectively slows the comparison and computations often required by data users.

In addition, inflation and currency fluctuation are not taken into account when calculating disaster-related damages. At present, estimating the monetary value of disasters is far from precise. Multi-standard reporting makes estimations difficult, as does the lack of standardisation of estimate components: for example, one estimation may include only damages to livestock, crops and infrastructure, while another may also include the cost of human lives lost. It is not always clear whether estimations are based on the cost of replacement or on the original value. Insurance figures, while using standard methodology, include only those assets that have been insured, which in most developing countries represent a minor proportion of the losses. A standard methodology for the estimation of the economic damages is urgently required to justify prevention and preparedness programmes.

Finally, assembling disaster-impact information faces the same problems as data reporting in general. To improve the quality of data, it is essential that a standard protocol for reporting procedures be established and followed by the main actors in disaster relief. Without such an agreement, data will always remain contradictory and incomplete.

Categories and definitions

Natural disasters are divided into the following categories: drought/ famine; earthquake; flood; high wind (cyclone, hurricane, storm, typhoon); landslide; volcano; and other (avalanche, cold wave, epidemic, food shortage, heat wave, tsunami).

Man-made disasters are defined as: accidents (transport accident, structural collapse); technological accidents (chemical, nuclear and mine explosions; chemical, atmospheric and oil pollution); and fire (forest and bush fires, as well as those caused by man).

Regional definitions are:

Africa: sub-Saharan countries and North Africa; America: North, Central and South America, as well as the Caribbean region; Asia: West, East , South and South-East Asia; and Europe: Western, Southern, Northern and Eastern (including the Russian Federation and the Republics of the former Soviet Union).

Disaster information from four different sources is reported regularly to CRED, and the register updated daily. The sources are: UNDHA situation reports; International Federation situation reports; *Lloyd's Casualty Week* (a bulletin published by Lloyd's of London with information on weather events, earthquakes, volcanic eruptions and different types of accidents worldwide. The primary sources for the bulletin are Reuters and Lloyd's country agents); and some World Wide Web pages on the Internet which provide daily news on disasters in specific regions (for example, the Pan-American Health Organization page).

Each new event is entered with date, type of disaster and country. Data on human or economic impact are consolidated at CRED at three-month intervals the first year. Annual updating is undertaken the following year.

Information is cross-checked each year, and occasionally new events added, using data from reinsurance companies (such as Munich-Reinsurance and Swiss-Reinsurance), the World Meteorological Organization, the UN Economic and Social Commission for Asia and the Pacific, as well as articles in specialised journals and unpublished university research. Finally, the annual list of disasters is sent for control to a focal point in countries affected by the disasters.

General comments

CRED's tables have been presented in the *World Disasters Report* since its inception in 1993. Some general comments should be made about the evolution of data collection.

The increase in the number of victims does not necessarily mean that human impact is increasing, but may simply be a reflection of better reporting. Information systems have improved vastly in the last 25 years and statistical data as a result are much more easily available.

After a 1992 workshop organised by CRED in Brussels, it was proposed that information on the number of homeless be reported separately from the population affected. In this report, both the numbers of people affected, and those made homeless, are given by disaster. But, to obtain a more realistic idea of the numbers involved, the number of homeless for the last five years should be added to the number of people affected.

Changes in national boundaries can cause ambiguities in the data, most notably the break-up of the Soviet Union and Yugoslavia, and the unification of Germany. In such cases, no attempt has been made to retrospectively desegregate or combine data. Statistics are presented for the country as it existed at the time the data were recorded.

In Table 7, figures for the annual average number of people reported killed or affected have been obtained as follows:

● New republics of the former Soviet Union: total killed or affected divided by four (they have been independent for four years);

● Former Soviet Union: total killed or affected divided by 21;

● East Germany: total killed or affected divided by 20 (reunification in 1989);

● Asian countries, especially India and China, are affected each year by different types of disasters. It is often difficult to estimate the correct number of victims, as the same disaster may affect several states, and information sources may not base their calculations on the same criteria. The information needs to be better authenticated in such situations.

The economic impact of natural disasters

The criteria used to determine the economic impact of disasters on a given country tend to take into account the degree to which that country's economy can absorb the damage caused by disaster.

At present, it is difficult to develop internationally acceptable standards. In spite of the serious lack of necessary information, it was decided to develop an indication by extrapolating from existing information.

The overall effect of a disaster depends, among other factors, on a state's economy. One disaster may have a catastrophic effect on a country with a weak economy; whereas the same disaster in another country with a stronger economy might not be considered serious at all.

Restricting their attention to the last decade, CRED's researchers decided to focus on the average estimated damage caused by a given disaster in a given country in relation to its annual gross national product (GNP). The economic impact indicator reflects the proportion of annual GNP lost due to disaster-related damage. When insufficient data are available to calculate this indicator, an estimated damage per person affected (in US dollars) is calculated by dividing the number of people affected in the disaster by the total damage figure reported for a similar disaster in the same country.

The number of people affected in the disaster are then multiplied by the amount per person to arrive at a figure for the total estimated damage caused, which is then used to compute the indicator as a proportion of GNP. The average estimated damage is calculated for each disaster which has affected a country during a one-year period; the sum of all important disasters will give the annual average estimated damage which is then presented as a percentage of one year's total annual GNP.

The use of the proportion of GNP lost due to disaster-related damage to indicate the importance of a disaster's economic impact on a given country seems to be satisfactory. However, it is important to standardise data collection on a national level, as the validity of the indicators is determined by the quality of data presented.

Tables 12 to 14

Refugee statistics tell vivid stories of persecution and human rights abuse. They also embody more ambiguous patterns of political upheaval, transition and flux. As such, governments cannot always be trusted to give full and unbiased accounts of refugee movements.

One country's refugee is another's illegal alien. Today's internally displaced person may be tomorrow's refugee. The term is sometimes a matter of law and policy, but just as frequently it is a matter of judgement.

The statistics in Tables 12, 13 and 14 represent the best judgements of the US Committee for Refugees, which has supplied the data, as to the number of uprooted people in two general categories: refugees and asylum seekers, and internally displaced people.

Tables 15 to 17

The data for these tables have been supplied by the Conflict Data Project of the Department of Peace and Conflict Research, Uppsala University,

Data sources

Tables 1 to 11
CRED
Catholic University of Louvain
School of Public Health
Clos Chapelle aux Champs
30-34
1200 Brussels
Belgium
Tel: (32)(2) 764 3327
Fax: (32)(2) 764 3328
Email: misson@epid.ucl.ac.be

Tables 12 to 14
US Committee for Refugees
1717 Massachusetts Ave NW
Suite 701
Washington DC 20036
USA
Tel: (1)(202) 347 3507
Fax: (1)(202) 347 3418
Email: uscr@irsa-uscr.org

Tables 15 to 17
Department of Peace and
Conflict Research
Uppsala University
Box 514
S-751 20 Uppsala
Sweden
Tel: (46)(18) 182352
Fax: (46)(18) 695102
Email:
Peter.Wallensteen@pcr.uu.se

Table 18
Development Assistance
Committee
OECD
2 Rue André Pascal
75775 Paris Cedex 16
France
Tel: (33)(1) 4524 8980
Fax: (33)(1) 4524 1650

Table 19
Food Aid Information Group
World Food Programme
426 Via Cristoforo Colombo
00145 Rome
Italy
Tel: (39)(6) 5228 2796
Fax: (39)(6) 5228 2451
Email: simon@wfp.org

Sweden. The Department was established in 1971 to conduct research and offer courses in peace and conflict studies. Research activities cover two specific areas: the origins and dynamics of conflict; and conflict resolution and international security issues.

The data presented here are based on at least two independent sources, if possible, drawn from publicly available material – newspapers, journals, research reports, government sources, etc. The Conflict Data Project evaluates the material, and occasionally consults outside experts. In exceptional cases, the Project makes its own estimates of, for example, battle-related deaths, building on information of intensity of fighting, weapons used, etc. All data are revised annually as new information becomes available.

A major armed conflict is defined as:

● A conflict where a government is an actor on one side of the conflict, facing an organised armed opponent (either another government or an organised faction). Data therefore cover situations of organised violence, not spontaneous violence, and situations where two actors face each other, i.e. conflicts, but not massacres by one armed actor against unarmed civilians;

● A conflict where at least 1,000 battle-related deaths have been recorded during the course of the conflict. This means that only conflicts approaching a common notion of "war" or "protracted conflict" are included. There are many minor conflicts in addition to those reported here. The battle-related deaths include targeted victims of the fighting, whether civilians or military. Other effects of battle are not included, e.g., starvation as a result of a society's break-down; and

● A conflict concerning government and/or territorial control, i.e., political in nature. Conflicts between criminal groups for example are not included, as the purpose is clearly economic gain only. It also means that mutinies are not included, unless they demand the replacement of an incumbent government.

(In the following tables, some figures may not match totals due to rounding.)

TABLES 1 and 2: Human impact by region. Annual average over 25 years (1970-1994)

Disasters with a natural trigger

	AFRICA	AMERICAS	ASIA	EUROPE	OCEANIA	TOTAL
Killed	76,485	8,988	55,922	2,240	94	143,728
Injured	1,017	15,180	37,288	3,475	135	57,096
Affected	11,450,827	4,481,691	111,473,882	561,580	* 653,827	128,621,807
Homeless	256,871	308,359	4,334,807	64,965	14,077	4,979,080
Total	11,785,200	4,814,218	115,901,899	632,260	668,133	133,801,711

Injured

Killed

Homeless

Affected

Disasters with a non-natural trigger

	AFRICA	AMERICAS	ASIA	EUROPE	OCEANIA	TOTAL
Killed	** 590	3,775	2,326	906	20	7,617
Injured	256	1,044	5,712	535	486	8,033
Affected	3,526	49,363	41,575	9,099	* 11,410	114,973
Homeless	2,384	1,734	6,359	7,784	64	18,325
Total	6,756	55,916	55,972	18,324	11,980	148,948

Killed

Injured

Homeless

Affected

* These figures have increased significantly from the data presented in 1995 due to detailed and complete reporting from Australia.

** The difference between figures reported in the *World Disasters Report 1995* and this year's figures for Africa is due to the cleaning of the data to remove conflict-related deaths and injuries which were included in the figures presented in 1995.

Worldwide disasters, excluding war, kill over 150,000 people a year and impact on the lives of 128 million. While figures for individual countries vary tremendously from year to year, the general trend is upwards. More people affected, more people killed. Africa has the most disaster deaths; Asia the largest number affected.

TABLES 3 and 4: Human impact by type. Annual average over 25 years (1970-1994)

Disasters with a natural trigger

	EARTHQUAKE	DROUGHT & FAMINE	FLOOD	HIGH WIND	LANDSLIDE	VOLCANO	TOTAL
Killed	21,593	73,606	12,361	28,194	1,560	1,014	138,329
Injured	30,952	n.a.	17,910	7,668	247	280	57,056
Affected	1,768,695	58,622,156	52,543,433	11,107,110	137,613	94,030	124,273,037
Homeless	232,406	22,720	3,502,014	1,111,092	107,434	14,764	4,990,430
Total	2,053,646	58,718,482	56,075,718	12,254,064	246,854	110,088	129,458,852

   Injured   Killed   Homeless   Affected

Disasters with a non-natural trigger

	ACCIDENT	TECHNOLOGICAL ACCIDENT	FIRES	TOTAL
Killed	3,667	617	3,333	7,617
Injured	1,701	5,583	751	8,035
Affected	17,290	53,558	44,125	114,973
Homeless	* 868	* 8,517	8,939	18,325
Total	23,526	68,275	57,148	148,950

   Killed   Injured   Homeless   Affected

* The difference between these two figures is due to the fact that accidents (e.g., train accidents) do not generally cause any homelessness; but, in the case of technological accidents, the authorities are often obliged to evacuate people from the immediate area of the disaster.

Drought and floods are the two most devastating natural phenomena causing disasters. Each affect well over 50 million people per year but, proportionally, they kill comparatively few. People caught in earthquakes or high-wind disasters run a far greater risk of death than those in flooded areas. The figures for technological accidents and fires, while still insignificant compared with natural disasters, are on the increase. Responding to technological disasters may well be a skill humanitarian agencies will have to develop in the future.

TABLES 5 and 6: Number of events by global region and type over 25 years (1970-1994)

Disasters with a natural trigger

	AFRICA	AMERICAS	ASIA	EUROPE	OCEANIA	TOTAL
Earthquake	40	129	234	163	85	651
Drought & Famine	272	51	88	15	15	441
Flood	168	373	628	138	139	1,446
Landslide	12	87	96	21	10	226
High Wind	84	428	683	214	200	1,609
Volcano	8	31	45	16	5	105
Other	218	90	190	95	7	600
Total	802	1,189	1,964	662	461	5,078

Volcano Landslide Drought, Famine Other Earthquake Flood High Wind

Disasters with a non-natural trigger

	AFRICA	AMERICAS	ASIA	EUROPE	OCEANIA	TOTAL
Accident	222	327	707	309	21	1,586
Technological accident	25	103	107	97	4	336
Fire	36	122	233	177	31	599
Total	283	552	1,047	583	56	2,521

Technological accident Fire Accident

Asia, partly because of its size but also because of the occurrence of most forms of natural hazards and an increasing number of technological hazards, is the continent most frequently affected by disaster. Globally, high-wind disasters are the most frequently reported, followed by accidents (mostly large transport accidents). This reflects their actual common occurrence, but also the ease with which they are identified and reported. Such disasters are very point specific, having a definite beginning and end. Others, like drought-related famine, are far more difficult to define.

TABLES 7 and 8: Total number of events by global region and type for 1995

Disasters with a natural trigger

	AFRICA	AMERICAS	ASIA	EUROPE	OCEANIA	TOTAL
Earthquake	1	8	11	3	0	23
Drought & Famine	6	1	3	1	0	11
Flood	16	17	37	17	0	87
Landlside	0	4	5	1	0	10
High Wind	1	24	20	4	0	49
Volcano	1	2	0	0	1	4
Other	2	13	8	5	1	29
Total	27	69	84	31	2	213

Volcano Landslide Drought, Famine Other Earthquake Flood High Wind

Disasters with a non-natural trigger

	AFRICA	AMERICAS	ASIA	EUROPE	OCEANIA	TOTAL
Accident	10	13	20	18	1	62
Technological accident	1	1	4	2	0	8
Fire	1	3	12	3	0	19
Total	12	17	36	23	1	89

Technological accident Fire Accident

The overall number of natural-trigger disasters reported for 1995 is up on the 1994 figure by some 30 per cent. For the Americas, the figure has risen by 80 per cent and for Europe by 48 per cent. Figures for disasters with a non-natural trigger, on the other hand, are underreported, as the source for much of this information are reinsurance companies, who do not make their annual reports until June.

TABLE 9: Annual average number of people reported killed or affected by disasters by country over 25 years (1970-1994)

Country	Killed	Affected	Country	Killed	Affected	Country	Killed	Affected
Africa			Haiti	168	219,861	Singapore	1	n.a.
Ethiopia	48,464	2,712,757	El Salvador	119	64,661	Macao	0	80
Sudan	6,115	981,061	Dominican Rep.	84	102,566			
Mozambique	4,547	1,255,159	Chile	62	165,225	**Europe**		
Somalia	906	37,003	Canada	62	19,895	Soviet Union	1,461	61,512
Nigeria	470	138,418	Venezuela	60	5,179	Russian Federat.	850	17,161
Chad	285	295,005	Bolivia	50	166,117	Tajikistan	544	21,275
Cameroon	142	40,749	Puerto Rico	47	160	Turkey	517	32,869
Algeria	139	28,108	Argentina	43	498,711	Estonia	227	35
Angola	138	128,488	Guyana	36	10,859	Italy	193	75,588
Zambia	123	103,121	Cuba	33	65,335	Spain	148	32,924
Egypt	121	7,666	Jamaica	19	54,187	Georgia	109	26,500
Kenya	109	401,723	Panama	13	7,020	France	88	33,412
Mauritania	97	271,131	Costa Rica	7	10,032	Romania	87	58,230
South Africa	94	262,926	Suriname	7	n.a.	United Kingdom	83	558
Niger	84	313,285	Bahamas	4	n.a.	Greece	67	29,072
Zaire	75	32944	Paraguay	3	17,703	Ukraine	66	102,333
Mali	74	209,027	Dominica	2	3,600	Yugoslavia	53	15,154
Tanzania	72	140,669	St. Lucia	2	2,944	Kyrgyzstan	42	48,750
Malawi	69	579,561	Martinique	2	1,060	Finland	39	n.a.
Burkina Faso	55	27,8845	Anguilla	1	n.a.	Serbia	35	n.a.
Madagascar	52	293,149	Belize	1	3,731	Poland	32	856
Guinea	39	2,204	Bermuda	1	n.a.	Uzbekistan	24	12,500
Uganda	34	59,405	Uruguay	1	948	Portugal	24	1,607
Zimbabwe	28	184,210	Trinidad & Tobago	0	2,000	Norway	23	n.a.
Swaziland	26	62,049	Guadeloupe	0	3,000	Armenia	23	52,000
Sierra Leone	25	520	Barbados	0	8	Azerbaijan	21	n.a.
Tunisia	24	7,567	Antigua	0	3,000	Germany, F. Rep.	18	4,220
Liberia	22	129	St. Vincent	0	1,726	Belgium	17	65
Congo	20	n.a.	**Asia**			Denmark	17	4
Morocco	19	27,675	Bangladesh	31,870	10,867,802	Ireland	17	140
Benin	18	121,968	China, P. Rep.	12,847	29,423,116	Albania	16	141,400
Senegal	17	291,256	India	4,728	63,723,255	Kazakhstan	13	7,500
Ghana	16	501,041	Iran	2,958	71,870	Hungary	12	n.a.
Burundi	15	288	Philippines	2,159	2,299,494	Moldova	12	6,250
Côte d'Ivoire	13	340	Indonesia	650	319,140	German Dem. Rep.	12	600
Libyan Arab Jam.	12	n.a.	Pakistan	575	1,023,233	Sweden	10	n.a.
Rwanda	12	164,145	Afghanistan	489	73,311	Switzerland	10	290
Djibouti	11	36,490	Nepal	365	251,819	Azores	10	852
Guinea Bissau	11	839	Viet Nam	316	1,480,076	Macedonia	8	400
Gambia	8	29,400	Japan	245	141,403	Czechoslovakia	8	4
Togo	8	24,465	Korea, Rep. of	174	66,732	Bulgaria	6	n.a.
Botswana	8	170,842	Thailand	158	538,204	Austria	4	n.a.
Sao Tomé & Princ.	7	7,483	Burma	107	241,067	Netherlands	3	480
Mauritius	7	39,526	Saudi Arabia	95	n.a.	Iceland	0	208
Namibia	5	10,000	Yemen Arab Rep.	86	121,000	Turkmenistan	0	75
Reunion	5	6,728	Sri Lanka	73	558,508	Belarus	0	10,000
Gabon	4	405	Taiwan	50	1,321			
Cen. African Rep.	3	573	Iraq	50	20,006	**Oceania**		
Comoros	3	15,418	Korea, Dem. Rep.	42	1,778	Papua New Guinea	41	13,024
Cape Verde Is.	3	304	Hong Kong	32	1,703	Australia	39	573,364
Lesotho	2	34,020	Laos	32	191,960	Solomon Islands	15	8,887
Eritrea	1	253	Cambodia	31	41,616	Fiji	10	45,598
Eq. Guinea	0	13	Yemen, P. D. Rep.	30	31,680	Vanuatu	3	6,244
			Malaysia	28	14,905	New Zealand	2	1,958
			Maldives	9	462	Samoa	1	n.a.
Americas			Lebanon	7	2,060	Western Samoa	0	10,280
Peru	4,160	517,194	United Arab Emir.	6	n.a.	Kiribati	0	42
Nicaragua	3,340	59,287	Mongolia	6	4,000	French Polynesia	0	200
Colombia	1,188	239,716	Oman	5	200	Tonga	0	5,192
Guatemala	978	156,440	Bahrain	5	n.a.	Tuvalu	0	40
Mexico	621	91,535	Israel	4	16	Wallis & Futuna	0	180
USA	523	45,778	Syria	4	5,367	Cook Islands	0	80
Honduras	476	50,952	Bhutan	2	2,623	New Caledonia	0	80
Brazil	409	1,878,252	Jordan	1	749	Tokelau	0	68
Ecuador	180	60,372						

Comments on Table 9

Africa
Egypt: Average number of affected has doubled since the *World Disasters Report 1995* because of a flood in 1994.
Kenya: The high number of affected is due to an epidemic of malaria.
Madagascar: The average number of affected people reflects the devastation of cyclones Daisy and Geralda in 1994.
Guinea: The country suffered in 1994 from a cholera epidemic which affected 24,000 people.
Djibouti: The 1994 flood which affected 100,000 people was one of the most important events during the 25-year period.

Americas
USA: The high number of victims is due to floods, storms and snowstorms in 1994.
Haiti: An important flood and the storm Gordon hit Haiti in 1994.

Asia
China: The average number of affected people shows an increase of 6 million since the last *Report*. This is due to floods, high winds and a major drought, and also to better reporting.
Philippines: The Philippines is one of the most disaster-prone countries of the region. High winds (Teresa and Welling typhoons), floods and the Mount Pinatuba volcanic eruption all contributed to the large numbers of people killed and affected in 1994.

Europe
The break-up of the former Yugoslavia, the former Soviet Union and the reunification of Germany during the reporting period mean that some states appear twice. Tajikistan, for instance, is included in the Soviet Union figures for the period up to its independence, thereafter it is recorded separately.

Oceania
Papua New Guinea: The volcanic eruption in 1994, which affected more than 100,000 people, dominates the 25-year average figures.

TABLES 10 and 11: Estimated annual average damage by region and by type over five years (1990-1994) in thousands US$

Disasters with a natural trigger

	AFRICA	AMERICAS	ASIA	EUROPE	OCEANIA	TOTAL
Earthquake	281,200	26,866,200	22,673,732	895,900	255,000	50,972,032
Drought & Famine	n.a.	1,781,000	69,955	12,188,600	1,008,400	15,047,955
Flood	424,755	19,275,665	81,689,348	81,495,345	11,100	182,896,213
Landslide	n.a.	25,400	217,700	60,100	n.a.	303,200
High Wind	398,165	38,515,052	33,929,641	104,784,160	1,207,700	178,834,718
Volcano	n.a.	10,000	220,888	n.a.	400,000	630,888
Other	47,000	4,538,401	9,556,300	978,999	0	15,120,700
Total	1,151,120	91,011,718	148,357,564	200,403,104	2,882,200	443,805,706

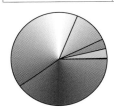

Volcano   Landslide   Drought, Famine   Other   Earthquake   Flood   High Wind

Disasters with a non-natural trigger

	AFRICA	AMERICAS	ASIA	EUROPE	OCEANIA	TOTAL
Accident	155,300	9,655,803	62,527,300	987,800	38,500	73,364,703
Technological accident	27,500	769,055	275,117	2,514,500	38,000	3,624,172
Fire	37,000	3,563,800	6,647,065	4,012,700	150,000	14,410,565
Total	219,800	13,988,658	69,449,482	7,515,000	226,500	91,399,440

Technological accident   Fire   Accident

Estimating the financial cost of disasters is extremely difficult. Most figures reported concentrate on capital and infrastructure losses. Therefore, the effect of disaster is under-reported in less developed countries, where losses have more to do with lost family income. Even so, total losses from disasters, excluding war losses, amounts to over $500 billion a year. Natural-trigger disaster losses are highest in Europe, not because of more disasters, but because more property is insured than in Asia, for example.

TABLE 12: Refugees and asylum seekers by country of origin

	1990	1991	1992	1993	1994	1995
AFRICA	5,414,900	5,321,500	5,730,600	5,812,400	5,857,650	5,285,650
Angola	435,700	443,200	404,200	335,000	344,000	313,000
Burundi	186,200	208,500	184,000	780,000	330,000	290,000
Chad	34,400	34,800	24,000	33,400	29,000	290,000
Djibouti	—	—	—	7,000	10,000	10,000
Eritrea	*	*	*	373,000	384,500	362,100
Ethiopia	1,066,300	752,400	834,800	232,200	190,750	110,750
Liberia	729,800	661,700	599,200	701,000	784,000	724,000
Mali	21,400	53,000	81,000	87,000	115,000	90,000
Mauritania	60,100	66,000	65,000	79,000	75,000	80,000
Mozambique	1,427,500	1,483,500	1,725,000	1,332,000	325,000	100,000
Niger	3,500	500	5,000	6,000	20,000	20,000
Rwanda	203,900	203,900	201,500	275,000	1,715,000	1,545,000
Senegal	24,400	27,600	15,000	18,000	17,000	22,000
Sierra Leone	—	181,000	200,000	260,000	260,000	364,000
Somalia	454,600	717,600	864,800	491,200	457,400	450,800
South Africa	40,000	23,700	11,100	10,600	—	—
Sudan	499,100	202,500	263,000	373,000	510,000	535,000
Togo	—	15,000	6,000	240,000	140,000	95,000
Uganda	12,300	14,900	15,100	20,000	15,000	10,000
Western Sahara	165,000	165,000	165,000	80,000	80,000	80,000
Zaire	50,700	66,700	66,900	79,000	56,000	56,000
EAST ASIA AND PACIFIC	698,900	811,450	502,000	797,400	690,050	640,850
Burma	50,800	112,000	86,700	289,500	203,300	160,300
Cambodia	344,500	392,700	148,600	35,500	30,250	26,200
China (Tibet)	114,000	114,000	128,000	133,000	139,000	141,000
Indonesia	—	6,900	5,500	9,400	9,700	9,500
Laos	67,400	63,200	43,300	26,500	12,900	8,900
Viet Nam	122,200	122,650	89,900	303,500	294,900	294,950
SOUTH AND CENTRAL ASIA	6,330,100	6,900,800	4,715,400	3,899,050	3,319,200	2,699,500
Afghanistan	6,027,100	6,600,800	4,286,000	3,429,800	2,835,300	2,220,200
Bangladesh	75,000	65,000	50,000	53,500	48,300	48,300
Bhutan	—	25,000	95,400	105,100	116,600	118,600
Sri Lanka	228,000	210,000	181,000	106,650	104,000	94,000
Tajikistan	*	*	52,000	153,000	165,000	170,400
Uzbekistan	*	*	51,000	51,000	50,000	48,000
MIDDLE EAST	3,554,400	2,792,500	2,849,300	2,975,000	3,826,950	3,951,000
Iran	211,100	50,000	65,400	39,000	54,250	54,600
Iraq	529,700	217,500	125,900	134,700	635,900	630,700
Kuwait	385,500	—	—	—	—	—
Palestine	2,428,100	2,525,000	2,658,000	2,801,300	3,136,800	3,265,700
EUROPE	0	120,000	2,529,800	1,952,650	1,775,800	1,823,400
Armenia	*	*	202,000	200,000	229,000	185,000
Azerbaijan	*	*	350,000	290,000	374,000	390,000
Bosnia-Herzegovina#	*	*	n.a.	n.a.	863,300	915,800
Croatia #	*	*	n.a.	n.a.	136,900	200,000
Georgia	*	*	130,000	143,000	106,800	105,000
Moldova	*	*	80'000	—	—	—
Turkey	—	—	—	—	13,000	17,000
Yugoslavia #	—	120,000	1,767,800	1,319,650	52,800	10,600
AMERICAS AND CARIBBEAN	152,400	118,250	104,250	97,500	120,550	79,450
Colombia	3,000	4,000	—	450	100	200
Cuba	2,900	1,400	1,650	1,400	30,600	4,000
El Salvador	37,200	24,200	22,800	21,900	16,200	15,850
Guatemala	57,400	46,700	45,750	49,200	45,050	35,450
Haiti	—	6,950	1,600	1,500	5,850	1,500
Nicaragua	41,900	25,400	30,850	23,050	22,750	22,450
Suriname	10,000	9,600	1,600			
WORLD TOTAL	16,150,700	16,064,500	16,431,350	15,534,000	15,590,200	14,479,850

Notes: — indicates zero or near zero; * country did not exist as of reporting date; n.a. not available, or reported estimates unreliable; # for 1992-93, refugees from Croatia and Bosnia included in Yugoslavia total, for 1994-95, Yugoslavia total includes only refugees from Serbia and Montenegro. Source: US Committee for Refugees.

TABLE 13: Refugees and asylum seekers by host country

	1990	1991	1992	1993	1994	1995
AFRICA	5,442,450	5,339,950	5,697,650	5,824,700	5,879,700	5,318,100
Algeria	189,400	204,000	210,000	121,000	130,000	120,000
Angola	11,900	10,400	9,000	11,000	11,000	10,300
Benin	800	15,100	4,300	120,000	50,000	25,000
Botswana	1,000	1,400	500	500	—	—
Burkina Faso	300	400	6,300	6,000	30,000	20,000
Burundi	90,700	107,000	107,350	110,000	165,000	140,000
Cameroon	6,900	6,900	1,500	2,500	2,000	2,000
Central African Republic	6,300	9,000	18,000	41,000	42,000	42,000
Congo	3,400	3,400	9,400	13,000	16,000	15,000
Côte d'Ivoire	270,500	240,400	195,500	250,000	320,000	290,000
Djibouti	67,400	120,000	96,000	60,000	60,000	25,000
Egypt	37,800	7,750	10,650	11,000	10,700	10,400
Ethiopia	783,000	534,000	416,000	156,000	250,000	308,000
Gabon	800	800	200	200	—	1,000
Gambia	800	1,500	3,300	2,000	1,000	6,000
Ghana	8,000	6,150	12,100	133,000	110,000	85,000
Guinea	325,000	566,000	485,000	570,000	580,000	640,000
Guinea-Bissau	1,600	4,600	12,000	16,000	16,000	20,000
Kenya	14,400	107,150	422,900	332,000	257,000	225,000
Lesotho	1,000	300	200	100	—	—
Liberia	—	12,000	100,000	110,000	100,000	120,000
Malawi	909,000	950,000	1,070,000	700,000	70,000	5,000
Mali	10,600	13,500	10,000	13,000	15,000	15,000
Mauritania	22,000	40,000	40,000	46,000	55,000	35,000
Mozambique	700	500	250	—	—	—
Namibia	25,000	30,200	150	5,000	1,000	1,000
Niger	800	1,400	3,600	3,000	3,000	17,000
Nigeria	5,300	4,600	2,900	4,400	5,000	6,000
Rwanda	21,500	32,500	24,500	370,000	—	—
Senegal	55,300	53,100	55,100	66,000	60,000	68,000
Sierra Leone	125,000	17,200	7,600	15,000	20,000	15,000
Somalia	358,500	35,000	10,000	—	—	—
South Africa	201,000	201,000	250,000	300,000	200,000	90,000
Sudan	726,500	717,200	750,500	633,000	550,000	470,000
Swaziland	47,200	47,200	52,000	57,000	—	—
Tanzania	266,200	251,100	257,800	479,500	752,000	703,000
Togo	—	450	350	—	5,000	10,000
Uganda	156,000	165,450	179,600	257,000	323,000	321,000
Zaire	370,900	482,300	442,400	452,000	1,527,000	1,332,000
Zambia	133,950	140,500	155,700	158,500	123,000	125,400
Zimbabwe	186,000	198,500	265,000	200,000	20,000	—
EAST ASIA AND PACIFIC	592,100	688,500	398,600	467,600	444,100	449,650
Australia	n.a.	23,000	24,000	2,950	5,300	4,700
China	5,000	14,200	12,500	296,900	297,100	294,100
Hong Kong	52,000	60,000	45,300	3,550	1,900	1,900
Indonesia	20,500	18,700	15,600	2,400	250	—
Japan	800	900	700	950	7,350	9,800
Korea	200	200	150	—	—	—
Macau	200	100	—	—	—	—
Malaysia	14,600	12,700	16,700	8,150	6,100	5,300
Papua New Guinea	8,000	6,700	3,800	7,700	9,700	9,500
Philippines	19,600	18,000	5,600	1,700	250	150
Singapore	150	150	100	—	—	—
Solomon Islands	—	—	—	—	3,000	1,000
Taiwan	150	150	150	—	—	—
Thailand	454,200	512,700	255,000	108,300	83,050	98,200
Viet Nam	16,700	21,000	19,000	35,000	30,100	25,0000
EUROPE	627,400	578,400	3,210,400	2,542,100	2,421,500	2,533,800
Armenia	*	*	300,000	290,000	295,800	304,000
Austria	22,800	27,300	82,100	77,700	59,000	55,900
Azerbaijan	*	*	246,000	251,000	279,000	238,000
Belarus	*	*	3,700	10,400	18,800	7,000
Belgium	13,000	15,200	19,100	32,900	19,400	16,400
Bulgaria	—	—	—	—	900	1,300
Croatia	*	*	420,000	280,000	188,000	189,500

	1990	1991	1992	1993	1994	1995
Czechoslovakia	1,600	2,800	2,200	*	*	*
Czech Republic	*	*	*	6,300	4,700	2,400
Denmark	5,500	4,600	13,900	23,300	24,750	9,600
Estonia	*	*	—	—	100	—
Finland	2,700	2,100	3,500	3,700	850	850
France	56,000	46,800	29,400	30,900	32,600	30,000
Germany	193,100	256,100	536,000	529,100	430,000	442,700
Greece	6,200	2,700	1,900	800	1,300	1,200
Hungary	18,300	5,200	40,000	10,000	11,200	9,100
Italy	4,800	31,400	19,100	33,550	31,800	61,400
Latvia	*	*	—	—	150	150
Lithuania	*	*	*			400
Macedonia	*	*	32,700	12,100	8,200	7,000
Netherlands	21,200	21,600	24,600	35,400	52,600	39,300
Norway	3,900	4,600	5,700	14,200	11,600	11,200
Poland	—	2,500	1,500	600	500	800
Portugal	100	200	—	2,250	600	500
Romania	—	500		1,000	600	1,300
Russian Federation	*	*	460,000	347,500	451,000	500,000
Slovak Republic	*	*	*	1,900	2,000	1,600
Slovenia	*	*	68,900	38,000	29,000	24,000
Spain	6,800	8,100	12,700	14,000	14,500	5,800
Sweden	28,900	27,300	88,400	58,800	61,000	12,300
Switzerland	37,000	41,600	81,700	27,000	23,900	27,000
Turkey	178,000	31,500	31,700	24,600	30,650	33,100
Ukraine	*	*	40,000	—	5,000	6,000
United Kingdom	25,000	44,700	24,600	28,100	32,000	44,000
Yugoslavia #	2,500	1,600	621,000	357,000	300,000	450,000
AMERICAS AND CARIBBEAN	229,050	218,800	249,000	272,450	297,300	260,750
Argentina	1,800	1,800	—	—	—	—
Belize	6,200	12,000	8,700	8,900	8,800	8,700
Bolivia	100	100	—	600	600	600
Brazil	200	200	200	1,000	2,000	2,000
Canada	36,600	30,500	37,700	20,500	22,000	25,000
Chile	—	—	—	100	200	300
Colombia	700	700	400	400	400	400
Costa Rica	26,900	24,300	34,350	24,800	24,600	24,600
Cuba	3,000	1,100	1,100	—	—	—
Dominican Republic	—	—	—	1,300	1,350	900
Ecuador	3,750	4,200	200	100	100	100
El Salvador	600	250	250	150	150	150
French Guiana	10,000	9,600	1,600	—	—	—
Guatemala	6,700	8,300	4,900	4,700	4,700	4,700
Honduras	2,700	2,050	150	100	100	50
Mexico	53,000	48,500	47,300	52,000	47,700	38,100
Nicaragua	500	2,800	5,850	4,750	300	450
Panama	1,200	1,300	850	950	900	800
Peru	600	600	400	400	700	700
United States	73,600	68,800	103,700	150,400	181,700	152,200
Venezuela	900	1,700	1,350	1,300	1,000	1,000
MIDDLE EAST	5,698,600	5,770,200	5,586,850	4,923,800	5,447,750	5,362,950
Bahrain	7,500	—	—	—	—	—
Gaza Strip	496,300	528,700	560,200	603,000	644,000	683,600
Iran	2,860,000	3,150,000	2,781,800	1,995,000	2,220,000	1,931,000
Iraq	60,000	48,000	64,600	39,500	120,500	124,900
Jordan	929,100	960,200	1,010,850	1,073,600	1,232,150	1,288,800
Kuwait	—	—	—	—	25,000	55,000
Lebanon	306,400	314,200	322,900	329,000	338,200	348,100
Oman	3,000	—	—	—	—	—
Saudi Arabia	300,000	34,000	27,400	25,000	17,000	13,200
Syria	280,700	293,900	307,500	319,200	332,900	377,600
United Arab Emirates	40,000	—	—	—	150	400
West Bank	414,300	430,100	459,100	479,000	504,000	517,400
Yemen	1,300	11,100	52,500	60,500	13,850	22,950
SOUTH AND CENTRAL ASIA	4,098,600	4,050,750	2,341,700	2,151,400	1,776,450	1,387,900
Afghanistan	—	—	52,000	35,000	20,000	18,400
Bangladesh	—	30,150	245,300	199,000	116,200	55,200

	1990	1991	1992	1993	1994	1995
India	415,800	402,600	378,000	325,600	327,850	320,600
Kazakhstan	*	*	—	6,500	300	6,500
Kyrgyzstan	*	*	—	3,500	350	7,600
Nepal	14,000	24,000	89,400	99,100	104,600	106,600
Pakistan	3,668,800	3,594,000	1,577,000	1,482,300	1,202,650	867,500
Tajikistan	*	*	—	400	2,500	2,500
Uzbekistan	*	*	—	—	2,000	3,000
World total	**16,688,200**	**16,646,60**	**17,484,200**	**16,182,05**	**16,266,800**	**15,313,150**

Notes: — indicates zero or near zero; * country did not exist as of reporting date; n.a. not available, or reported estimates unreliable; # beginning in 1992, includes only Serbia and Montenegro. Source: US Committee for Refugees.

Table 12 reflects the number of "refugees and asylum seekers in need of protection and/or assistance" by country of origin. Often, these are populations officially designated as refugees by host governments, UNHCR and the UN Relief and Works Agency for Palestine Refugees in the Near East. Others may lack such official recognition, but in USCR's opinion warrant aid and/or protection based on the persecution they suffered or their fear of return, and on the precarious nature of their asylum. This table does not include refugees who have found a durable solution to their plight, such as permanent resettlement. As asylum states do not always specify countries of refugee origin, Table 12 may understate the total number of refugees and asylum seekers from a given country.

Table 13 is a retabulation of Table 12, listing refugees and asylum seekers by host country. Some, mainly in Europe and North America, have entered into asylum proceedings during the past year.

TABLE 14: Significant populations of internally displaced people

	1990	1991	1992	1993	1994	1995
AFRICA	13,504,000	14,222,000	17,395,000	16,890,000	15,730,000	11,885,000
Angola	704,000	827,000	900,000	2,000,000	2,000,000	1,500,000
Burundi	—	—	—	500,000	400,000	300,000
Djibouti	—	—	—	140,000	50,000	—
Eritrea	*	*	*	200,000	—	—
Ethiopia	1,000,000	1,000,000	600,000	500,000	400,000	200,000
Ghana	—	—	—	—	20,000	150,000
Kenya	—	—	45,000	300,000	210,000	210,000
Liberia	500,000	500,000	600,000	1,000,000	1,100,000	1,000,000
Mozambique	2,000,000	2,000,000	3,500,000	2,000,000	500,000	500,000
Rwanda	—	100,000	350,000	300,000	1,200,000	500,000
Sierra Leone	—	145,000	200,000	400,000	700,000	1,000,000
Somalia	400,000	500,000	2,000,000	700,000	500,000	300,000
South Africa	4,100,000	4,100,000	4,100,000	4,000,000	4,000,000	2,000,000
Sudan	4,500,000	4,750,000	5,000,000	4,000,000	4,000,000	4,000,000
Togo	—	—	—	150,000	100,000	—
Uganda	300,000	300,000	—	—	—	—
Zaire	—	—	100,000	700,000	550,000	225,000
AMERICAS AND CARIBBEAN	1,126,000	1,221,000	1,354,000	1,400,000	1,400,000	1,280,000
Colombia	50,000	150,000	300,000	300,000	600,000	600,000
El Salvador	400,000	150,000	154,000	—	—	—
Guatemala	100,000	150,000	150,000	200,000	200,000	200,000
Haiti	—	200,000	250,000	300,000	—	—
Honduras	22,000	7,000	—	—	—	—
Nicaragua	354,000	354,000	—	—	—	—
Panama	0	10,000	—	—	—	—
Peru	200,000	200,000	500,000	600,000	600,000	480,000
EAST ASIA AND PACIFIC	340,000	680,000	699,000	595,000	613,000	555,000
Burma	200,000	500,000	500,000	500,000	500,000	500,000
Cambodia	140,000	180,000	199,000	95,000	113,000	55,000
SOUTH AND CENTRAL ASIA	3,085,000	2,685,000	1,810,000	880,000	1,775,000	1,550,000
Afghanistan	2,000,000	2,000,000	530,000	n.a.	1,000,000	450,000
India	85,000	85,000	280,000	250,000	250,000	250,000
Sri Lanka	1,000,000	600,000	600,000	600,000	525,000	850,000
Tajikistan	*	*	400,000	30,000	—	—
EUROPE	1,048,000	1,755,000	1,626,000	2,765,000	5,195,000	4,945,000
Azerbaijan	*	*	216,000	600,000	630,000	670,000
Bosnia-Herzegovina	*	*	740,000	1,300,000	1,300,000	1,300,000
Croatia	*	*	340,000	350,000	290,000	180,000
Cyprus	268,000	268,000	265,000	265,000	265,000	265,000
Georgia	*	*	15,000	250,000	260,000	280,000
Moldova	*	*	20,000	—	—	—
Russian Federation	*	*	—	n.a.	450,000	250,000
Soviet Union	750,000	900,000	*	*	*	*
Turkey	30,000	30,000	30,000	n.a.	2,000,000	2,000,000
Yugoslavia #	—	557,000	—	—	—	—
MIDDLE EAST	1,300,000	1,450,000	800,000	1,960,000	1,710,000	1,400,000
Iran	—	—	—	260,000	—	—
Iraq	500,000	700,000	400,000	1,000,000	1,000,000	1,000,000
Lebanon	800,000	750,000	400,000	700,000	600,000	400,000
Yemen	—	—	—	—	110,000	—
WORLD TOTAL	20,403,000	22,013,000	23,684,000	24,490,000	26,423,000	21,615,000

Notes: — indicates zero or near zero; * country did not exist as of reporting date; n.a. not available, or reported estimates unreliable; # beginning in 1992, Yugoslavia includes only Serbia and Montenegro. Source: US Committee for Refugees.

> *Listed here are the most significant populations of "internally displaced people". Internally displaced people fled their homes for the same reasons as refugees, but did not cross an international border. Often, estimates of the number of the internally displaced are particularly speculative and should be viewed with a degree of caution.*

TABLE 15: Number of major armed conflicts by region per year over five years (1990-1995)

	1990	1991	1992	1993	1994	1995
Europe / Total	1	2	4	6	5	3
War	—	1	2	4	1	2
Intermediate	1	1	2	2	4	1
Middle East / Total	5	7	5	6	6	6
War	1	3	1	1	2	1
Intermediate	4	4	4	5	4	5
Asia / Total	15	12	13	11	11	12
War	6	7	7	4	2	2
Intermediate	9	5	6	7	9	10
Africa / Total	11	11	7	7	7	6
War	9	9	7	4	2	3
Intermediate	2	2	—	3	5	3
Americas / Total	4	4	3	3	3	3
War	3	1	3	2	—	—
Intermediate	1	3	—	1	3	3

A major armed conflict is defined as a conflict in which at least 1,000 battle-related deaths have been reported since the beginning of conflict. There are two categories of major armed conflict: "war" - in which more than 1,000 battle-related deaths were incurred during the year in question, and, "intermediate" armed conflicts in which less than 1,000 battle-related deaths were incurred during the year in question.

During the 1990s all regions of the world have witnessed a major armed conflict. Some subregions are exceptions, for example, in North America and Australasia there have been only minor armed conflicts, notably in Mexico and in Papua New Guinea. Northern and Western Europe is also an exception, where the Northern Ireland and Basque issues show a pattern of gradual decline.

Most conflicts occur in Asia and Africa, respectively, the largest regions in terms of population and territory. But the patterns are not as might be expected. There are actual declines in the number of armed conflicts as well as locations with serious armed conflict. By early 1996, the conflicts that had plagued Southern Africa were all ended. In the Horn of Africa, too, protracted conflicts had been brought to an end or were comparatively low level (e.g., Ethiopia, Eritrea, Somalia and Sudan). In Asia, the subregion of South-East Asia displayed a pattern of fewer and/or less violent conflicts.

In Europe, however, the dissolution of Yugoslavia and the Soviet Union was accompanied by a series of major armed conflicts, i.e. in Croatia, Bosnia-Herzegovina, Georgia (Abkhazia), Azerbaijan (Nagorno-Karabakh) and Chechnya. The conflict in Chechnya was the only major armed conflict not contained by early 1996.

The world as a whole was one of contradicting tendencies: conflicts ending in some parts of the world, conflicts augmenting in others.

TABLE 16: Number of battle-related deaths in major armed conflicts per region per year over five years (1990-1995)

	1990	1991	1992	1993	1994	1995
Europe	74	6,000 -10,000	11,200 - 21,400	14,200 - 42,000	1,500	1,000 - 33,000
Middle East	>3,500	>16,000	3,300 - 4,500	3,000 - 4,000	4,800 - 12,000	3,250 - 5,500
Asia	>15,000	>16,000	14,000 - 60,000	23,500 - 35,000	6,300 - 15,000	>6,200
Africa	33,500	37,000	14,000 - 40,000	25,500	25,000 - 35,000	15,000
Americas	6,000 - 7,500	3,200 - 6,200	>5,400	<3,400	<1,400	<1,700

This table, with numbers of battle-related deaths by region from 1990 to 1995, shows some interesting phenomena. For the whole period, Africa and Asia have seen the most battle-related deaths, with more people killed than in the more highly-publicised wars in Europe (e.g., the former Yugoslavia) or in the Middle East (e.g., the Gulf war). Several factors may contribute to this: the widespread use of land-mines, and the lower availability of humanitarian assistance and medical aid might be an explanation, reflecting a structural difference between wars in poor and in rich regions.

Nevertheless, some regional differences are important. The situation in Europe, for example, has changed dramatically. From the low-level conflict in Northern Ireland, the continent has seen two devastating conflicts: in the former Yugoslavia (particularly in 1992 and 1993) and in Chechnya (1995). In both these conflicts, regular army units have battled, entailing considerable destruction. In the Middle East, the technical warfare of the Gulf war (1991) and the war in Yemen (1994) was particularly brutal. For Asia, the continuing nightmare of Afghanistan was a major problem throughout the period (with a peak in 1992), as was the conflict in Sri Lanka. In Africa, many different conflicts led to a high number of casualties, i.e., Ethiopia (from 1990 to 1991), Somalia (1991), Angola (from 1990 to 1995), Liberia (in 1990 and in 1995), Sudan (1992) and Algeria (since 1993). For Central and South America, there has been a steady reduction in battle-related deaths.

Because this table refers only to battle-related deaths, it does not include figures for Rwanda's genocide.

Table 17: Number of major armed conflicts by type of incompatibility per region per year over five years (1990-1995)

	1990		1991		1992		1993		1994		1995	
	G	T	G	T	G	T	G	T	G	T	G	T
Europe	—	1	—	2	—	4	—	6	—	5	—	3
Middle East	1	4	2	5	2	3	2	4	2	4	2	4
Asia	5	10	3	8	4	9	4	7	4	7	4	8
Africa	8	3	8	3	6	1	6	1	6	1	5	1
Americas	4	—	4	—	3	—	3	—	3	—	3	—

Major armed conflicts are caused by two types of incompatibilities: Government (G) concerning type of political system, a change of central government or its composition, or, Territory (T) concerning control of territory (interstate conflict), secession or autonomy.

This table shows that the issues at stake vary between regions. In Europe, all major armed conflicts have concerned territorial issues, whereas in the Americas, control over central government has been at issue. In other regions the patterns are more varied. It is interesting to observe that in ethnically-divided Africa, most conflicts still concern control of government, rather than the deliberate breaking-up of existing states.

A clearly discernible overall trend can be found in the data. More and more frequently, conflicts concern the viability of states. The issues of conflict are more often territorial: certain parts of a state wishing to break away, or at least become more autonomous. There have also been challenges to the existing central state, but where such challenges failed and resulted in a fragmentation of the state, no actor controlled the whole country. The first tendency could be seen in the cases of the former Yugoslavia, Georgia, Azerbaijan, the Russian Federation (Chechnya), and Myanmar. The second was most clearly witnessed in the case of Somalia, Liberia, Sierra Leone and Afghanistan and (earlier) Lebanon. Rwanda was a special case, where the attempted genocide of one group by another resulted in the former taking control and the latter fleeing the country.

The overall trend may have deeper causes. The end of the Cold War seemed associated with a general decline in state authority, caused by several circumstances. First, Cold War dynamics propped up a number of governments and states, by providing military and financial support which ceased with the end of the Cold War. Second, the post-Cold War period has been marked by a strong belief in market mechanisms. This has further undermined states and governments, as they were no longer the providers of economic goods to the population. As the tax base was small, and economic development weak, states were weakened in the face of opponents. Thus, a pattern of reduction of state authority - not replaced by a strong civil society - was visible in Africa, the Middle East and Central Asia.

TABLE 18: Non-food emergency and distress relief, grant disbursements in millions US$ over ten years (1985-1994)

Donor	1985	1986	1987	1988	1989	1990	1991	1992	1993	1994
Australia	8.81	5.90	17.31	7.94	6.78	12.23	13.23	29.56	26.56	25.49
Austria	3.36	3.03	5.92	14.07	22.68	43.96	93.87	145.83	123.45	127.04
Belgium	1.92	2.12	1.21	1.75	1.59	4.59	5.71	13.18	19.05	14.02
Canada	54.66	25.88	26.44	55.86	29.69	45.76	85.14	78.86	248.54	227.51
Denmark	0.00	0.00	0.00	0.00	0.00	108.27	52.83	104.85	77.14	78.62
Finland	5.03	9.81	26.21	19.74	31.96	70.54	102.24	61.55	21.61	27.48
France	0.00	0.00	0.00	0.00	0.00	0.00	0.00	25.88	125.08	122.23
Germany	18.01	22.44	28.16	35.82	30.68	45.21	415.31	680.32	549.52	392.53
Ireland	1.32	1.08	1.06	1.22	1.33	2.09	2.89	2.10	5.15	8.53
Italy	90.66	188.44	124.88	145.21	84.15	104.06	456.33	137.39	341.69	105.40
Japan	6.94	1.87	2.22	8.90	19.61	26.46	20.48	14.93	40.37	31.08
Luxembourg	0.00	0.00	0.40	1.50	2.00	3.80	10.30	7.21	8.49	5.09
Netherlands	22.92	25.37	28.96	33.99	24.40	63.58	109.74	197.45	303.29	302.37
New Zealand	0.91	0.20	1.24	0.61	0.00	3.85	1.51	5.12	4.96	2.68
Norway	21.03	22.44	20.37	41.74	50.38	88.62	77.60	86.48	113.21	180.75
Portugal	0.00	0.00	0.00	0.00	0.00	0.00	0.11	0.11	8.35	3.70
Spain	0.00	0.00	0.00	0.00	1.20	5.00	8.42	6.43	7.74	4.74
Sweden	78.90	99.34	133.29	110.28	214.50	124.49	181.65	342.56	277.28	334.17
Switzerland	19.14	25.90	64.53	40.87	46.49	46.75	67.78	68.61	66.85	80.98
United Kingdom	43.03	27.21	21.49	32.36	31.72	37.95	116.48	56.83	187.27	260.52
United States	225.00	193.00	183.00	170.00	210.00	221.00	596.00	521.00	669.00	1,132.00
Total	**601.64**	**654.03**	**686.69**	**721.86**	**809.16**	**1,058.21**	**2,417.62**	**2,586.25**	**3,224.60**	**3,466.93**
Proportion for refugees	**573.01**	**627.42**	**130.30**	**177.47**	**225.76**	**348.21**	**1,052.41**	**1,713.27**	**1,976.28**	**2,399.66**

Table 19: Breakdown of food-aid deliveries by category per year over nine years (1987-1995) in thousands tonnes - cereals in grain equivalent

	1987	1988	1989	1990	1991	1992	1993	1994	1995*
Emergency	1,987	3,281	2,389	2,767	3,540	4,991	4,202	4,214	3,798
Project	3,938	4,057	2,872	2,837	2,980	2,578	2,498	2,794	2,735
Programme	8,479	7,510	6,473	8,038	6,650	7,663,	10,170	5,587	3,167
World total	14,404	14,848	11,734	13,642	13,170	15,232	16,870	12,595	9,700

Table 18: The total funding going into humanitarian assistance from the OECD countries now amounts to over $3,400 million a year. However, the rapid rise in spending seen over the first half of this decade is unlikely to be continued over the second half. Figures for 1994 may well represent a high point of humanitarian spending. What is significant is that this rapid rise in funding seems to have been mostly directed at refugee-related operations, a field of activities which shows little sign of decreasing.

*Table 19: Total food-aid disbursements include both food aid for development and for humanitarian assistance. Today, however, humanitarian food aid takes up a greater proportion than ever of total food aid, as much as 70 per cent, according to some experts. * 1995 figures are provisional.*

Political relief: Disasters and humanitarian crises can and do affect every country of the world. Addressing the effects of crisis is the core business of humanitarian action, but almost always addressing the causes of crisis requires timely, consistent and determined action by governments or inter-governmental organisations. Humanitarian action alone rarely solves a crisis. Keeping humanitarian action independent while ensuring that root causes are addressed through political systems is the essence of good disaster response.

Displaced people, Bosnia, 1995. Sebastiao Salgado/Magnum

Putting humanitarian values up front

The International Conference of the Red Cross and Red Crescent is the only global forum that brings together governments from across the globe with the world's leading humanitarian movement to focus on the present practice and future direction of humanitarian assistance and protection.

For four days in December 1995, 142 governments – out of the 186 Parties to the Geneva Conventions – met with 164, of the 169 recognised, National Red Cross and Red Crescent Societies, the International Federation of Red Cross and Red Crescent Societies and the International Committee of the Red Cross (ICRC) to debate humanitarian issues in a humanitarian forum.

In discussing the entire business of the Movement, the conference covered almost every aspect of today's humanitarian catastrophes: assistance and protection for those caught up in war; the interface between humanitarian, political and military action, and those seeking refuge from it; the development of better practice in humanitarian assistance; and the strengthening of national institutions to respond to crises and need.

The conference agreed five resolutions, to which the Movement and the states present have put their names and should be held accountable for practising what they preach.

The full text of all the resolutions and background papers of the conference can be found on the International Federation's Internet Web page (http://www.ifrc.org).

The conference reviewed the practice of International Humanitarian Law, i.e., the Geneva Conventions and their Additional Protocols. This body of law puts limits on the means and methods of war, with the purpose of reducing the adverse effects of war on the victims, military and civilian, who do not or no longer take part in it.

The nature of warfare is changing but the principles of the Conventions hold true. The conference allowed governments to focus on issues such as the use of food or water as a weapon of war, the age at which children are recruited into armies and the ongoing campaign to ban land-mines.

Outside of the war zone, the conference made substantial progress clarifying the purpose of humanitarian assistance in today's complex emergencies. Humanitarian assistance has to be kept independent, neutral and free from political influence, but at the same time such assistance can only address the effects or symptoms of crisis, not its causes.

Addressing the causes is primarily the business of governments. They must do their job and give the humanitarian agencies the space needed to provide assistance and protection to war and disaster victims. Dealing with humanitarian needs in countries under sanctions provided a focus for this debate.

Building on this theme, the *Code of Conduct*, reported on in Chapter 13, was supported by all states present and they committed themselves to encourage non-governmental organisations in their countries to use it to improve working standards, policies and planning, and responsiveness to people's urgent needs.

In summary, the following are the decisions of the conference on protecting the victims of war:

Protection of civilians in conflict: The conference highlighted the spread of acts of genocide, "ethnic cleansing", widespread murder, forced displacement and the use of force to prevent people returning home, hostage-taking, torture, rape and arbitrary detention. The conference strongly condemned systematic and massive killings of civilians in armed conflicts and urged states and all parties to armed conflicts to comply in all circumstances, and ensure compliance by their armed forces, with the relevant principles and norms of International Humanitarian Law.

Women: The conference strongly condemned sexual violence which, in the conduct of armed conflict, can be seen as a war crime. There was strong lobbying for the establishment and strengthening of mechanisms to investigate, bring to justice and punish all those responsible.

Children: The conference strongly condemned the deliberate killing of children, sexual exploitation, abuse and violence against children in warfare. Recruitment and conscription of children under 15 into the armed forces or armed groups was strongly condemned.

Famine and war: The conference strongly condemned attempts to starve civilian populations in armed conflicts and urged parties to conflict to

Box 12.1 What does International Humanitarian Law say?

International Humanitarian Law is a complicated body of law which includes the Geneva Conventions and their Additional Protocols. It is a body of law to which most states in the world have subscribed and provides vital protection for those caught up in warfare. Its key principles can be summarised in seven points.

People *hors de combat* and those who do not take a direct part in hostilities are entitled to respect for their lives and their moral and physical integrity. They shall in all circumstances be protected and treated humanely without any adverse distinction.

It is forbidden to kill or injure an enemy who surrenders or who is *hors de combat*.

The wounded and sick shall be collected and cared for by the party to the conflict which has them in its power. Protection also covers medical personnel, establishments, transports and equipment. The emblem of the red cross or the red crescent is the sign of such protection and must be respected.

Captured combatants and civilians under the authority of an adverse party are entitled to respect for their lives, dignity, personal rights and convictions.

They shall be protected against all acts of violence and reprisals. They shall have the right to correspond with their families and to receive relief.

Everyone shall be entitled to benefit from fundamental judicial guarantees. No one shall be held responsible for an act he has not committed. No one shall be subjected to physical or mental torture, corporal punishment or cruel or degrading treatment.

Parties to a conflict and members of their armed forces do not have an unlimited choice of methods and means of warfare. It is prohibited to employ weapons or methods of warfare of a nature to cause unnecessary losses or excessive suffering.

Parties to a conflict shall at all times distinguish between the civilian population and combatants in order to spare civilian population and property. Neither the civilian population as such nor civilian persons shall be the object of attack. Attacks shall be directed solely against military objectives. ∎

Sources, references, further information

International Museum of the Red Cross and Red Crescent. *The humanitarian endeavour: from Solferino (1859) to Sarejevo (1995)*. Geneva, 1995

The International Committee of the Red Cross and the International Federation of Red Cross and Red Crescent Societies. *International Red Cross and Red Crescent Handbook*. 13th Edition, Geneva, 1994.

Macalister-Smith, P. *International humanitarian assistance*. Dordrecht: Martinus Nijhoff, 1985.

International Federation on WWW: http://www.ifrc.org

maintain conditions in which the civilian population would be able to provide for its own needs.

Reunification of families: The conference demanded that all parties to armed conflict avoid any action aimed at, or having the effect of, causing the separation of families contrary to International Humanitarian Law.

Water and war: The conference called upon parties to conflict to take all feasible precautions to avoid acts liable to destroy or damage water sources and systems of water supply, purification and distribution solely or primarily used by civilians, and not to hinder access by the civilian population to water.

Anti-personnel land-mines: The conference urged all states and competent organisations to increase support for mine-clearance efforts in affected states.

Laser weapons: The conference welcomed adoption by a Review Conference of States on conventional weapons of a new Fourth Protocol on blinding laser weapons as an important step in the development of International Humanitarian Law.

In summary, the following are the decisions of the conference on humanitarian assistance and protection:

Internally displaced persons and refugees: The conference reminded states of their International Humanitarian Law commitments in this connection and called on them to ensure adequate access and resources for humanitarian organisations to those in need.

Developmental relief: The conference called on states to explore ways of encouraging a developmental approach to relief through their humanitarian assistance programming.

Code of Conduct: The conference welcomed the *Code of Conduct* and invited all states and National Societies to encourage non-governmental organisations to both abide by the principles and spirit of the *Code* and consider registering their support.

Sanctions: The conference encouraged states to consider the possible negative impact of sanctions on the humanitarian situation of civilians in a targeted state and of third states that may be adversely affected. States were also called on to permit relief operations of a strictly humanitarian character for the benefit of the most vulnerable civilian groups.

Independence of humanitarian action: The conference called on states to recognise the need for the Movement to maintain a clear separation between its humanitarian action and actions of a political, military or economic nature carried out by governments, intergovernmental bodies and other agencies during humanitarian crises.

National Societies' independence: The conference reaffirmed the mandate of National Societies as autonomous humanitarian organisations auxiliary to their governments, while recognising the need for National Societies to maintain independence and autonomy.

Providing service: States were called on to use National Societies as cost-effective providers of health care, social services and emergency assistance for the most vulnerable.

Building capacity: The conference called upon the Federation and the ICRC, in cooperation with National Societies, to draw up a model law of recognition of a National Society. ∎

Securing standards: Values, standards and ethics must be at the core of all humanitarian work, helping agencies and their staff as far more is expected of them. In the growing complexity of today's crises, in which war, economic collapse, grave violations of human rights and natural catastrophes pile up one upon the other, relief workers need clear and consistent guidance to ensure a strong ethical basis for their work. Without this basis, agreeing more practical and even technical standards for relief delivery is difficult.

Camp guards, Rwanda, 1994. James Nachtwey/Magnum

Governments back global standards

In December 1995 142 governments, represented in Geneva at the International Conference of the Red Cross and Red Crescent, unanimously gave their support to a *Code of Conduct* for relief workers.

The *Code of Conduct for the International Red Cross and Red Crescent Movement and NGOs in Disaster Relief* was developed and agreed upon by eight of the world's largest disaster-response agencies in the summer of 1994, i.e., Caritas Internationalis, Catholic Relief Services, International Federation of Red Cross and Red Crescent Societies, International Save the Children Alliance, Lutheran World Federation, Oxfam, World Council of Churches (all members of the Steering Committee for Humanitarian Response) and the International Committee of the Red Cross (ICRC).

The *Code* represents a significant initiative in setting standards for disaster response and is being used by the International Federation, the ICRC and the other founding agencies to monitor their own standards of relief delivery and to encourage other agencies to set similar standards.

Since its publication, the *Code* has been endorsed by a fast-growing number of aid agencies operating internationally. The full list of agencies is given below.

Throughout the 1980s and 1990s, there has been a steady growth in the number of non-governmental organisations (NGOs), both national and international, involved in disaster relief. In the autumn of 1994 there were over 120 NGOs registered in Kigali, the war-ravaged capital of Rwanda. With the implementation of the Dayton Peace Accords, NGOs, big and small, were flocking into Bosnia in early 1996.

Many of these agencies, including National Red Cross and Red Crescent Societies, the church agencies, Oxfam, Save the Children Fund and CARE, have a history going back many decades and have gained a reputation for effective work.

Others, more recently formed, such as *Médecins sans Frontières*, have rapidly evolved to become respected operators. Along with these large and well-known agencies, there is today a multitude of small, newly formed groups, often coming into existence to assist in one specific disaster or in a specialised field of work.

For all these agencies, from the old to the new, from multi-million dollar outfits to one-man organisations, there is no accepted body of professional standards to guide their work. It is still assumed in many countries that disaster relief is essentially "charitable" work and therefore anything done in the name of helping disaster victims is acceptable.

Agencies, whether experienced or newly created, can make mistakes, be misguided and sometimes deliberately misuse the trust that is put in them. Disaster relief is no longer limited to individual gestures. The Federation alone assisted some 13 million disaster victims in 1995 with an overall relief expenditure reaching 327 million Swiss francs for the year.

The immediacy of disaster relief can often lead NGOs to unwittingly put pressure on themselves, which leads to short-sighted and inappropriate work: programmes which rely on foreign imports or expertise, projects which pay little attention to local custom and culture, and activities which accept the easy and high media-profile tasks of relief but leave for others the less appealing and more difficult ones of disaster preparedness and long-term rehabilitation.

External pressures

All NGOs, big and small, are susceptible to these internal and external pressures. As they are required to do more and as the incidence of complex disasters involving natural, economic and often military factors increases, the need for some sort of basic professional code becomes more and more imperative.

It is for all these reasons that six of the world's oldest and largest networks of NGOs came together in 1994 with the International Red Cross and Red Crescent Movement to draw up a professional *Code of Conduct* to set for the first time universal basic standards to govern the way they should work in disaster assistance.

The *Code of Conduct*, like most professional codes, is a voluntary one. It is applicable to any international or national voluntary humanitarian organisation.

It lays down ten points of principle to which all NGOs and the Red Cross and Red Crescent should adhere in their disaster-response work and describes the relationships that agencies working in disasters should seek with donor governments, host governments and the United Nations (UN) system.

The *Code* is self-policing; one NGO is not going to force another to act in a certain way and there is as yet no international association for disaster-response NGOs which possesses any authority to sanction its members. It is hoped that NGOs around the world will find the *Code* useful and want to commit themselves publicly to abide by it.

Governments and donor bodies may want to use the *Code* as a yardstick to judge the conduct of those agencies with which they work. And disaster-affected communities have a right to expect that those who seek to assist them measure up to these standards.

The Federation has undertaken to keep a register of all agencies who commit themselves to abiding by the principles of the *Code*.

The *Code* also sets out recommendations for governments and intergovernmental organisations to help facilitate the effective participation and coordination of the Movement and NGOs in disaster response.

Above all the *Code* reminds governments that the Movement and humanitarian NGOs act from humanitarian motives and need the support of governments in respecting their independence and impartiality.

This respect can be turned into action by assistance from governments to ensure that the Movement and NGOs have rapid and impartial access to disaster victims, by facilitating the flow of relief goods to disaster victims through waiving commercial import restrictions and by inviting the Movement and NGOs to join relief coordinating mechanisms.

Code of Conduct for the International Red Cross and Red Crescent Movement and NGOs in Disaster Relief: Principal Commitments

The full text of the *Code* can be found in the *World Disasters Report 1994* in English, French, Spanish, Arabic, Finnish and Japanese, and on the Federation's Internet Web site (http://www.ifrc.org). The ten principal points that signatories have agreed to abide by are given below:

1. The humanitarian imperative comes first.
2. Aid is given regardless of race, creed or nationality of the recipients and without adverse distinction of any kind. Aid priorities are calculated on the basis of the need alone.
3. Aid will not be used to further a particular political or religious standpoint.
4. We shall endeavour not to act as instruments of government foreign policy.
5. We shall respect culture and custom.
6. We shall attempt to build disaster response on local capacities.
7. Ways shall be found to involve programme beneficiaries in the management of relief aid.
8. Relief aid must strive to reduce future vulnerabilities to disaster as well as meeting basic needs.
9. We hold ourselves accountable to both those we seek to assist and those from whom we accept resources.
10. In our information, publicity and advertising activities, we shall recognise disaster victims as dignified humans, not hopeless objects.

The *Code of Conduct* Register

The Federation is keeping a public record of all NGOs who register their commitment to the *Code of Conduct,* and will publish the list periodically in the *World Disasters Report*. The full text of the *Code*, including a registration form, is published by the Federation as a short booklet in English, French, Spanish and Arabic, and is available upon request. A 10-minute, *Code of Conduct* training video with notes is also available – in the same four languages as the booklet – at a small charge.

For details, write to: Disaster and Refugee Policy Department, International Federation of Red Cross and Red Crescent Societies, PO Box 372, 1211 Geneva 19, Switzerland, tel.: (41)(22)730 4222; fax: (41)(22)733 0395; email: walker@ifrc.org

In the *World Disasters Report 1995*, 46 agencies were listed as registering their commitment to the *Code.* That number has now reached 72, and the full list (as at 1 March 1996) is as follows:

Country	Agency name
Australia	Care Australia
Austria	Austrian Relief Programme (ARP)
Bangladesh	Youth Approach for Development and Cooperation (YADC)
Belgium	Agora – Vitrine du Monde
Belgium	Centre International de Formation des Cadres du Développement (C.I.F.C.D.)
Belgium	ICA – ZAGREB (Institute of Cultural Affairs International)
Belgium	OXFAM
Benin	Conseil des Activités Educatives du Bénin
Canada	Adventist Development and Relief Agency (ADRA)
Côte d'Ivoire	Adventist Development and Relief Agency (ADRA)

Country	Agency name
Croatia	ADEH International
Croatia	Pax Christi (Germany)
Denmark	Adventist Development and Relief Agency (ADRA)
Ethiopia	Selam Children's Village
France	Enfants Réfugiés du Monde
France	Enfants du Monde
Germany	Adventist Development and Relief Agency (ADRA)
Greece	Institute of International Social Affairs
India	Action for Social and Human Acme (ASHA)
India	Adventist Development and Relief Agency (ADRA)
India	Ambiha Charitable Trust
India	Mahila Udyamita Vikas Kalya Evan Siksha Sansthah
Italy	Associazione Amici dei Bambini
Italy	Caritas Internationalis
Italy	Centro Internazionale di Cooperazione allo Sviluppo (C.I.C.S.)
Italy	Comitato Collaborazione Medica (CCM)
Italy	Comitato di Coordinamento delle Organizzazioni per il Servizio Volontario
Italy	International College for Health Cooperation in Developing Countries (CUAMM)
Italy	Movimondo
Italy	Reggio Terzo Mondo (R.T.M.)
Italy	Volontari Italiani Solidarieta Paesi Emergenti (V.I.S.P.E.)
Laos	Adventist Development and Relief Agency (ADRA)
Lebanon	Disaster Control Centre
Luxembourg	Amicale Rwanda-Luxembourg
Myanmar	Adventist Development and Relief Agency (ADRA)
New Zealand	Tear Fund
Norway	Norwegian Organisation for Asylum Seekers
Norway	Norwegian Refugee Council
Philippines	Adventist Development and Relief Agency (ADRA)
Russia	Adventist Development and Relief Agency (ADRA)
Sri Lanka	Adventist Development and Relief Agency (ADRA)
Swaziland	Save the Children Fund
Sweden	Swedish Organisation for Individual Relief (SOIR)
Switzerland	Ananda Marga Universal Relief Team (AMURT)
Switzerland	Catholic Relief Services
Switzerland	Commission Internationale Catholique pour les Migrations
Switzerland	Food for the Hungry International
Switzerland	International Save the Children Alliance
Switzerland	International Committee of the Red Cross
Switzerland	International Federation of Red Cross and Red Crescent Societies
Switzerland	Lutheran World Federation
Switzerland	World Council of Churches
Switzerland	World Vision International
Thailand	Adventist Development and Relief Agency (ADRA)
UK	Adventist Development and Relief Agency (ADRA)
UK	Children in Crisis
UK	Feed the Children
UK	Human Appeal International
UK	International Extension College

Country	Agency name
UK	Marie Stopes International
UK	Medical Emergency Relief International (MERLIN)
UK	OXFAM
UK	Save the Children Fund
UK	United Kingdom Foundation for the Peoples of the South Pacific
USA	International Rescue Committee
USA	International Medical Corps
USA	Lutheran World Relief
USA	Operation USA
USA	Truck Aid International
USA	Women's Commission for Refugee Women and Children
USA	World Association of Girl Guides and Girl Scouts
Zaire	OXFAM

■

Global and local: Across the world there is a global humanitarian network larger and more experienced than any other which links together the 169 National Red Cross and Red Crescent Societies. Indigenous, independent, impartial, each National Society understands and serves the needs of disaster victims, the poor, the socially deprived, those most vulnerable. The daily action of the National Societies at the local level is the foundation upon which all the International Federation's assistance is built.

Aiding displaced people, Azerbaijan, 1994. Ian Berry/Magnum

WORLD DISASTERS REPORT 1996

Reaching out across the world

Contact details for the International Red Cross and Red Crescent Movement and National Societies.

THE INTERNATIONAL RED CROSS AND RED CRESCENT MOVEMENT

THE INTERNATIONAL FEDERATION OF RED CROSS AND RED CRESCENT SOCIETIES
P.O. Box 372
1211 Genève 19
SWITZERLAND

Tel.   (41)(22) 730 42 22
Fax   (41)(22) 733 03 95
Tlx   (045)412 133 FRC CH
Tlg.   LICROSS GENEVA
Eml   secretariat@ifrc.org

THE INTERNATIONAL COMMITTEE OF THE RED CROSS
19 avenue de la Paix
1202 Genève
SWITZERLAND

Tel.   (41)(22) 734 60 01
Fax   (41)(22) 733 20 57
Tlx   414 226 CCR CH
Tlg.   INTERCROIXROUGEGENE
Eml   icrc.gva@gwn.icrc.org

National Red Cross and Red Crescent Societies

(National Red Cross and Red Crescent Societies, listed alphabetically by International Organization for Standardization Codes for the Representation of Names of Countries, English spelling.)

Details correct as of 1 March 1996. Please forward any corrections to the Federation's Information Resource Centre in Geneva.

Afghan Red Crescent Society
P.O. Box 3066
Shahre - Naw Kabul
AFGHANISTAN

Tel.   32357 / 32211 / 34288
Tlx   arcs af 24318
Tlg.   SERAMIASHT KABUL

Albanian Red Cross
C.P. 1511
Tirana
ALBANIA

Tel.   (355)(42) 25855 / 22037
Fax   (355)(42) 25855 / 27508
Tlg.   ALBCROSS TIRANA

Algerian Red Crescent
15 bis, Boulevard Mohammed V
Alger
ALGERIA

Tel.   (213) 2645727 / 2645728 /
       2610741
Fax   (213) 2649787
Tlx   56056 HILAL ALGER
       ou 66442 CRA DZ
Tlg.   HILALAHMAR ALGER

Andorra Red Cross
Prat de la Creu 22
Andorra La Vella
ANDORRA

Tel.   (376) 825225
Fax   (376) 828630
Tlx   AND 208 "Att. CREU ROJA"

Angola Red Cross
Caixa Postal 927
Luanda
ANGOLA

Tel.   (244)(2) 336543 / 333991
Fax   (244)(2) 345065
Tlx   3394 CRUZVER AN

**Antigua and Barbuda Red
Cross Society**
P.O. Box 727
St. Johns, Antigua W.I.
ANTIGUA AND BARBUDA

Tel.   (1)(809) 4620800
Fax   (1)(809) 4620800
Tlx   2195 DISPREP
       "For Red   Cross" or
       2145 CWTXAGY
       "For Red Cross"

Argentine Red Cross
Hipolito Yrigoyen 2068
1089 Buenos Aires
ARGENTINA

Tel.   (54)(1) 9511391 / 9511854 /
       9512389
Fax   (54)(1) 9527715
Tlx   21061 CROJA AR
Tlg.   ARGENCROSS BUENOS
       AIRES

Armenian Red Cross Society
Antarain str. 188
Yerevan 375019
ARMENIA, REPUBLIC OF

Tel.   (7)(8852) 560630 /583630
       /564881
Fax   (7)(8852) 583630
Tlx   243345 ODER SU,
       Country code 64

Australian Red Cross
P.O. Box 100
East Melbourne, Vic. 3002
AUSTRALIA

Tel.   (61)(3) 94185200
Fax   (61)(3) 94190404
Tlx   34221 RECRO AA
Eml   kevins@vifp.monash.edu.au

Austrian Red Cross
Postfach 39
1041 Wien 4
AUSTRIA

Tel.   (43)(1) 58900-0
Fax   (43)(1) 58900-199
Tlx   oerk a 133111
Tlg.   AUSTROREDCROOS WIEN
Eml   ladislav@via.at

**Red Crescent Society of
Azerbaijan**
Prospekt Azerbaidjan
19 Baku
AZERBAIJAN

Tel.   (7)(8922) 931912 / 938481 /
       936346
Fax   (7)(8922) 931578

The Bahamas Red Cross Society
P.O. Box N-8331
Nassau
BAHAMAS

Tel.   (1)(809) 3237370 / 3237371 /
       3237372 / 3237373 / 3284415
Fax   (1)(809) 3237404
Tlx   20657 BAHREDCROSS
Tlg.   BAHREDCROSS NASSAU

Bahrain Red Crescent Society
P.O. Box 882
Manama
BAHRAIN

Tel.   (973) 293171
Fax   (973) 291797
Tlg.   HILAHAMAR MANAMA

**Bangladesh Red Crescent
Society**
G.P.O. Box 579
Dhaka
BANGLADESH

Tel.   (880)(2) 407908 / 406902 /
       400188 / 400189 / 402540
Fax   (880)(2) 831908
Tlx   632232 BDRC BJ
Tlg.   RED CRESCENT DHAKA

**The Barbados Red Cross
Society**
Red Cross House
Jemmotts Lane
Bridgetown
BARBADOS

Tel.   (1)(809) 4262052
Fax   (1)(809) 4262052
       "For Red Cross"
Tlx   2201 P.U.B. T.L.X. W.B.
Tlg.   REDCROSS BARBADOS

Red Cross Society of Belarus
35, Ulitsa Karla Marxa
220030 Minsk
BELARUS

Tel.   (375)(172) 271417
Fax   (375)(172) 272620
Tlx   252290 KREST SU

Belgian Red Cross
Ch. de Vleurgat 98
1050 Bruxelles
BELGIUM

Tel.   (32)(2) 6454411
Fax   Communauté francophone
       (32)(2) 6460439
       Communauté flamande
       (32)(2) 6460441
Tlx   24266 BELCRO B
Tlg.   CROIXROUGE BELGIQUE
       BRUXELLES
Eml   belgian.redcross@
       infoboard.be

Belize Red Cross Society
P.O. Box 413
Belize City
BELIZE

Tel.   (501)(2) 73319
Fax   (501)(2) 30998
Tlx   BTL BOOTH 211 Bze
      attn. Red Cross

Red Cross of Benin
B.P. No. 1
Porto-Novo
BENIN

Tel.   (229) 212886
Fax   (229) 214927
Tlx   -1131 CRBEN

Bolivian Red Cross
Casilla No. 741
La Paz
BOLIVIA

Tel.   (591)(2) 340948 / 326568 /
      376874
Fax   (591)(2) 359102
Tlx   2220 BOLCRUZ
Tlg.   CRUZROJA - LA PAZ

Botswana Red Cross Society
P.O. Box 485
Gaborone
BOTSWANA

Tel.   (267) 352465 / 312353
Fax   (267) 312352
Tlg.   THUSA GABORONE

Brazilian Red Cross
Praa Cruz Vermelha No. 10/12
20.230 Rio de Janeiro - RJ
BRAZIL

Tel.   (55)(21) 2323223 / 2326325
      (Direct line to President and
      Officers) / 2210252
      (Switchboard,
      day and night)
Fax   (55)(21) 2426760
Tlx   (38) 2130532 CVBR BR
Tlg.   BRAZCROSS RIO DE
      JANEIRO

Bulgarian Red Cross
61, Dondukov Boulevard
1527 Sofia
BULGARIA

Tel.   (359)(2) 441443 / 441444 /
      441445
Fax   (359)(2) 441759
Tlx   23248 B CH K BG
Tlg.   BULGAREDCROSS SOFIA

Burkinabé Red Cross Society
01 B.P. 340
Ouagadougou 01
BURKINA FASO

Tel.   (226) 300877
Fax   (226) 363121
Tlx   LSCR 5438 BF
      OUAGADOUGOU

Burundi Red Cross
B.P. 324
Bujumbura
BURUNDI

Tel.   (257) 223159 / 223576
Fax   (257) 23159
Tlx   5081 CAB PUB BDI
      5082 CAB PUB BDI

Cambodian Red Cross Society
17, Vithei de la Croix-Rouge
Cambodgienne
Phnom-Penh
CAMBODIA

Cameroon Red Cross Society
B.P. 631
Yaoundé
CAMEROON

Tel.   (237) 224177
Fax   (237) 224177
Tlx   (0970) 8884 KN

The Canadian Red Cross Society
1800 Alta Vista Drive
Ottawa
Ontario KIG 4J5
CANADA

Tel.   (1)(613) 7393000
Fax   (1)(613) 7311411
Tlx   CANCROSS 05-33784
Tlg.   CANCROSS OTTAWA
Eml   pwharram@redcross.ca

Red Cross of Cape Verde
Caixa Postal 119
Praia
CAPE VERDE

Tel.   (238) 611701 / 614169 /
      611621
Fax   (238) 614174 / 613909
Tlx   6004 CV CV

Central African Red Cross Society
B.P. 1428
Bangui
CENTRAL AFRICAN
REPUBLIC

Tel.   (236) 612223
Fax   (236) 613561 / 612223
Tlx   DIPLOMA 5213
      "Pour Croix-Rouge"

Red Cross of Chad
B.P. 449
N'djamena
CHAD

Tel.   (235) 515218 / 513434
Tlg.   CROIXROUGE
      N'DJAMENA

Chilean Red Cross
Correo 21, Casilla 246 V
Santiago de Chile
CHILE

Tel.   (56)(2) 7771448
Fax   (56)(2) 7370270
Tlx   340260 PBVTR CK
Tlg.   "CHILECRUZ"

Red Cross Society of China
53 Ganmian Hutong
100010 Beijing
CHINA

Tel.   (86)(1) 5124447 / 5135838
Fax   (86)(1) 512 4169
Tlx   210244 CHNRC CN
Tlg.   HONGHUI BEIJING

Colombian Red Cross Society
Apartado Aereo 11-10
Bogota, D.C.
COLOMBIA

Tel.   (57)(1) 2506611 / 2319445
       2319207 (Presidencia)
       2259775 (Internacionales)
Fax   (57)(1) 2319208 (Presidencia)
       2319006 (Internacionales)
       2319127 (Socorro Nacional)
Tlx   45433 CRC CO
Tlg.   CRUZ ROJA BOGOTA

Congolese Red Cross
B.P. 4145
Brazzaville
CONGO

Tel.   (242) 824410
Fax   (242) 828825
Tlx   UNISANTE 5364
       Pour "Croix-Rouge"

Costa Rican Red Cross
Apartado 1025
San José 1000
COSTA RICA

Tel.   (506) 2337033 / 2553761 /
       2553759
Fax   (506) 2237628
Tlx   2547 COSTACRUZ SAN JOS
Tlg.   COSTACRUZ SAN JOS

**Red Cross Society of Côte
d'Ivoire**
P.O. Box 1244
Abidjan 01
COTE D'IVOIRE

Tel.   (225) 321335
Fax   (225) 225355
Tlx   24122 SICOGI CI

Croatian Red Cross
Ulica Crvenog kriza 14
41000 Zagreb
CROATIA

Tel.   (385)(1) 415458/469
Fax   (385)(1) 450072
Eml   redcross@hck.hr

Cuban Red Cross
Calle Calzada No. 51 Vedado
Ciudad Habana
C.P. 10400
CUBA

Tel.   (53)(7) 324664 / 326005
Fax   (53)(7) 326005
Tlx   511149 MSP CU para Cruz
       Roja
Tlg.   CRUROCU HABANA

Czech Red Cross
Thunovska 18
CZ-118 04 Praha 1
CZECH REPUBLIC

Tel.   (42)(2) 245 10347   (Switch-
board)
       245 10318
       (Foreign Relations)
       245 10220
       (Tracing Department)
Fax   (42)(2) 245 10318 / 532340
Tlx   122 400 csrc c
Tlg.   CROIX PRAHA

Danish Red Cross
P.O. Box 2600
DK-2100 København O
DENMARK
Tel.   (45)(31) 381444
Fax   (45)(31) 421186
       383966 (International
       Department)
Tlx   15726 DANCRO DK
Tlg.   DANCROIX KØBENHAVN
Eml   redcross.dk@applelink.
       apple.com

**Red Crescent Society of
Djibouti**
B.P. 8
Djibouti
DJIBOUTI

Tel.   (253) 352451 / 353552
       (int. 14)
Fax   (253) 355049
Tlx   5871 PRESIDENCE DJ

Dominica Red Cross Society
National Headquarters
Federation Drive
Goodwill
Commonwealth of Dominica
DOMINICA
Tel.
Fax   (1)(809) 4487708
Tlx   8625 TELAGY DO - for
       Dominica RC
       (Only in case of emergency
       if fax and telegram do not
       work)
Tlg.   DOMCROSS

Dominican Red Cross
Apartado Postal 1293
Santo Domingo, D.N.
DOMINICAN REPUBLIC

Tel.   (1)(809) 6823793 / 6897344
Fax   (1)(809) 6822837
Tlx   rca sdg 4112
       "PARA CRUZ ROJA DOM"
Tlg.   CRUZ ROJA
DOMINICANA,
       SANTO DOMINGO

Ecuadorean Red Cross
Casilla 17-01-2119
Quito
ECUADOR

Tel.   (593)(2) 582485 (Presidencia)
       582479 (Direccion Tecnica)
       516089 (Difusion y
       Relaciones Pub)
       514587 (Administracion)
Fax   (593)(2) 570424
Tlx   CRUZRO 2662
Tlg.   CRUZ ROJA QUITO

Egyptian Red Crescent Society
29, El Galaa Street
Cairo
EGYPT

Tel.   (20)(2) 5750558 / 5750397
Fax   (20)(2) 5740450
Tlx   93249 ERCS UN
Tlg.   124 HELALHAMER

Salvadorean Red Cross Society
Apartado Postal 2672
San Salvador
EL SALVADOR

Tel. (503) 2227743 / 2227749
Fax (503) 2227758
Tlx 20550 cruzalva
Tlg. CRUZALVA SAN
SALVADOR

Red Cross of Equatorial Guinea
Apartado postal 460
Malabo
EQUATORIAL GUINEA

Tel. 2398
Tlx 099/1111 EG.PUB MBO
"Favor Transmetien Cruz Roja
Tel. 2393"

Estonia Red Cross
Lai Street 17
EE-0001 Tallinn
ESTONIA

Tel. (37)(22) 444265
Fax (37)(22) 441491
Tlx 173491

Ethiopian Red Cross Society
P.O. Box 195
Addis Ababa
ETHIOPIA

Tel. (251)(1) 449364 / 159074
Fax (251)(1) 512643
Tlx 21338 ERCS ET
Tlg. ETHIOCROSS
ADDISABABA
Eml ercs@padis.gn.apc.org

Fiji Red Cross Society
GPO Box 569
Suva
FIJI

Tel. (679) 31413 / 314138
Fax (679) 303818
Tlx 2279 Attn: Red Cross
(Public facility)
Tlg. REDCROSS SUVA

Finnish Red Cross
P.O. Box 168
FI-00141 Helsinki
FINLAND

Tel. (358)(0) 12931
(24-hour service)
Fax (358)(0) 654149
5801329
(Blood Transfusion Service)
Tlx 12 11331 FINCR SF
123130 SPRV SF
(Blood Transfusion
Service)
Tlg. FINCROSS HELSINKI
Eml kaloo@redcross.fi

French Red Cross
1, place Henry-Dunant
F-75384 Paris Cedex 08
FRANCE

Tel. (33)(1) 44431100 / 30211919
(24-hour emergency service)
44431375/76/77 (Operational
Centre)
Fax (33)(1) 44431101 / 44431388
(Operational Centre)
Tlx CR PARIS 642760 F CRPAR
Tlg. CROIROUGE PARIS 086

The Gambia Red Cross Society
P.O. Box 472
Banjul
GAMBIA

Tel. (220) 392405 / 393179 /
392347
Fax (220) 394921
Tlx 2338 REDCROSS GV
Tlg. GAMREDCROSS BANJUL

German Red Cross
Postfach 1460
53004 Bonn
GERMANY

Tel. (49)(228) 5411
Fax (49)(228) 541290
Tlx 886619 DKRB D
Tlg. DEUTSCHROTKREUZ
BONN
Eml drk@drk.de

Ghana Red Cross Society
P.O. Box 835
Accra
GHANA

Tel. (233)(21) 662298
Fax (233)(21) 662298
Tlx 2655 GRCS
Tlg. GHANACROSS ACCRA

Hellenic Red Cross
Rue Lycavittou 1
Athènes 106 72
GREECE

Tel. (30)(1) 3646005 / 3628648
3615606 / 3621681
Fax (30)(1) 3613564. 3615606
(International Dept.)
Tlx 225156 EES GR
Tlg. HELLECROIX ATHENES

Grenada Red Cross Society
P.O. Box 551
St. George's
GRENADA

Tel. (1)(809) 4401483
Fax (1)(809) 4401483

Guatemalan Red Cross
3a Calle 8 - 40, Zona 1
Guatemala, C.A.
GUATEMALA

Tel. (502)(2) 532026 / 532027
532028 / 26518
Tlx 5366 CROJA GU
Tlg. GUATECRUZ
GUATEMALA

Red Cross Society of Guinea
B.P. 376
Conakry
GUINEA

Tel. (224) 443825
Tlx 22101

**Red Cross Society of
Guinea-Bissau**
Caixa postal 514-1036 BIX,
Codex
Bissau
GUINEA-BISSAU

Tel. (245) 212405
Tlx 251 PCE BI

The Guyana Red Cross Society
P.O. Box 10524
Georgetown
GUYANA
Tel.   (592)(2) 65174
Fax   (592)(2) 66523
Tlx   2226 FERNA GY
       "For Guyana Red Cross"
Tlg.   GUYCROSS
       GEORGETOWN H

Haitian National Red Cross Society
CRH B.P. 1337
Port-au-Prince
HAITI

Tel.   (509) 225553 / 225554 /
       231035
Fax   (509) 231054
Tlg.   HAITICROSS PORT AU
       PRINCE

Honduran Red Cross
7a Calle
entre 1a. y 2a. Avenidas
Comayagela D.C.
HONDURAS

Tel.   (504) 378876 / 374628 /
       378654 / 372240
Fax   (504) 380185
Tlx   1437 CRUZ R HO
Tlg.   HONDUCRUZ
       COMAYAGUELA

Hungarian Red Cross
Magyar Vöröskereszt
1367 Budapest 5, Pf. 121
HUNGARY

Tel.   (36)(1) 1313950 / 1317711
       (International Department)
Fax   (36)(1) 1533988
Tlx   224943 REDCR H
Tlg.   REDCROSS BUDAPEST
Eml   vivas@slip.hrc.hu

Icelandic Red Cross
Postbox 5450
125 Reykjavik
ICELAND

Tel.   (354) 56267722
Fax   (354) 5623150
Eml   rki@centrum.is

Indian Red Cross Society
Red Cross Building
1 Red Cross Road
New Delhi 110001
INDIA

Tel.   (91)(11) 3716441/2/3
Fax   (91)(11) 3717454
Tlx   3166115 IRCS IN
Tlg.   INDCROSS NEW DELHI

Indonesian Red Cross Society
P.O. Box 2009
Jakarta
INDONESIA

Tel.   (62)(21) 7992325
Fax   (62)(21) 7995188
Tlx   66170 MB PMI IA
Tlg.   INDONCROSS JKT

Red Crescent Society of the Islamic Republic of Iran
Ostad Nejatolahi Avenue
Tehran
IRAN, ISLAMIC REPUBLIC OF

Tel.   (98)(21) 8849077 / 8849078
Fax   (98)(21) 8849079
Tlx   224259 RCIA-IR
Tlg.   CROISSANT-ROUGE
       TEHERAN

Iraqi Red Crescent Society
P.O. Box 6143
Baghdad
IRAQ

Tel.   (964)(1) 8862191 /5343922
Fax   (964)(1) 8840872
Tlx   213331 HELAL IK
Tlg.   REDCRESCENT BAGHDAD

Irish Red Cross Society
16, Merrion Square
Dublin 2
IRELAND

Tel.   (353)(1) 6765135 / 6765136
       6765137 / 6765679 / 6765686
Fax   (353)(1) 6767171 /6614461
Tlx   32746 IRCS EI
Tlg.   CROSDEARG DUBLIN

Italian Red Cross
12, Via Toscana
I - 00187 Roma
ITALY
Tel.   (39)(6) 47591
Fax   (39)(6) 4883541(Service
       Affaires Internationales) /
       44244534
Tlx   613421 CRIROM I
Tlg.   CRIROM 00187

Jamaica Red Cross
76 Arnold Road
Kingston 5, Jamaica W.I.
JAMAICA

Tel.   (1)(809) 9847860-3
Fax   (1)(809) 9848272
Tlx   COLYB JA 2397
       "For Red Cross"
       alternate
       ELECTRON JA 2212
       "For Red Cross"
Tlg.   JAMCROSS KINGSTON

Japanese Red Cross Society
1-1-3 Shiba Daimon, 1-Chome,
Minato-ku
Tokyo-105
JAPAN
Tel.   (81)(3) 34381311
Fax   (81)(3) 34358509
Tlx   JARCROSS J 22420
Tlg.   JAPANCROSS TOKYO
Eml   gbh11244@niftyserv.or.jp

Jordan National Red Crescent Society
P.O. Box 10001
Amman 11151
JORDAN

Tel.   (962)(6) 773141/773142/
       773687
Fax   (962)(6) 750815
Tlx   22500 HILAL JO
Tlg.   HALURDON AMMAN

Kenya Red Cross Society
P.O. Box 40712
Nairobi
KENYA

Tel.   (254)(2) 503781 / 503789 /
       503816
Fax   (254)(2) 503845
Tlx   25436 IFRC KE
Tlg.   KENREDCROSS NAIROBI

**Red Cross Society of the
Democratic People's Republic
of Korea**
Ryonwa 1, Central District
Pyongyang
KOREA, DEMOCRATIC
PEOPLE'S REPUBLIC OF

Tel.   (850)(2) 816048
Fax   (850)(2) 814644;
       814410/14/16
Tlx   5355 DAEMUN KP
Tlg.   KOREACROSS
       PYONGYANG

**The Republic of Korea
National Red Cross**
32 - 3ka, Namsan-dong
Choong-Ku
Seoul 100 - 043
KOREA, REPUBLIC OF

Tel.   ((2)(2) 7559301
Fax   (82)(2) 7740735 7 7520258
       (International Relations
       Department)
Tlx   ROKNRC K28585
Tlg.   KORCROSS SEOUL

Kuwait Red Crescent Society
P.O. Box 1359
Safat
KUWAIT

Tel.   (965) 4818086 / 4818085 /
       4815478
Fax   (965) 4839114
Tlx   22729

Lao Red Cross
B.P. 650
Vientiane
LAO PEOPLE'S DEMOCRATIC
REPUBLIC

Tel.   (856)(21) 216610 / 212036
       (Président)
       222398 (Secrétaire Général)
Fax   (856)(21) 215935
       c/o Représentation de la
       Fédération
Tlx   4491 TE via PTT LAOS
       c/o Représentation de la
       Fédération
Tlg.   CROIXLAO VIENTIANE

Latvian Red Cross
28, Skolas Street
Rigas
LV-1350
LATVIA

Tel.   (371)(2) 275635 / 274154 /
       275716 /275406 /
       (371)(7) 310902 (International)
Fax   (371)(2) 275635 / (371)(7)
       310902 (International)

Lebanese Red Cross
Rue Spears
Beyrouth
LEBANON

Tel.   (961)(1) 372801/2/3/4 /
       865561
Fax   (961)(1) 863299
Tlx   CROLIB 20593 LE
Tlg.   LIBACROSS BEYROUTH

Lesotho Red Cross Society
P.O. Box 366
Maseru 100
LESOTHO

Tel.   (256) 313911
Fax   (266) 310166
Tlx   4515 LECROS LO
Tlg.   LESCROSS MASERU
Eml   lrcs@wn.apc.org

Liberian Red Cross Society
P.O. Box 20-5081
1000 Monrovia 20
LIBERIA

Tel.   (231) 225172
Tlx   44210 / 44211 / 44214

Libyan Red Crescent
P.O. Box 541
Benghazi
LIBYAN ARAB JAMAHIRIYA

Tel.   (218)(61) 9095827 / 9099420
Fax   (218)(61) 9095829
Tlx   40341 HILAL PY
Tlg.   LIBHILAL BENGHAZI

Liechtenstein Red Cross
Heiligkreuz 25
FL-9490 Vaduz
LIECHTENSTEIN

Tel.   (41)(75) 2322294
Fax   (41)(75) 2322240
Tlg.   ROTESKREUZ VADUZ

Lithuanian Red Cross Society
Gedimino Ave 3a
2600 Vilnius
LITHUANIA

Tel.   (370)(2) 619923
Fax   (370)(2) 610437

Luxembourg Red Cross
B.P. 404
L - 2014 Luxembourg
LUXEMBOURG

Tel.   (352) 450202-1
Fax   (352) 457269
Tlg.   CROIXROUGE
       LUXEMBOURG

**The Red Cross of the Former
Yugoslav Republic of
Macedonia**
No. 13 Bul. Koco Racin
91000 Skopje
MACEDONIA, THE FORMER
YUGOSLAV REPUBLIC OF

Tel.   (389)(91) 114355
Fax   (389)(91) 230542

Malagasy Red Cross Society
B.P. 1168
101 Antananarivo
MADAGASCAR

Tel.   (261)(2) 22111
Fax   (261)(2) 35457
Tlx   22248 COLHOT MG
       "Pour Croix-Rouge"

Malawi Red Cross Society
P.O. Box 30096
Capital City
Lilongwe 3
MALAWI

Tel.   (265) 732877 / 732878
board
Fax   (265) 730210
Tlx   44276

Malaysian Red Crescent Society
JKR 32, Jalan Nipah
Off Jalan Ampang
55000 Kuala Lumpur
MALAYSIA

Tel.   (60)(3) 4578122. 4578236
       4578348 / 457815 /. 4578227
Fax   (60)(3) 4579867
Tlx   MACRES MA 30166
Tlg.   MALREDCRES KUALA
       LUMPUR
Eml   secgen@mrcs.po.my

Mali Red Cross
B.P. 280
Bamako
MALI

Tel.   (223) 224569
Fax   (223) 220414
Tlx   2611 MJ

Malta Red Cross Society
104 St Ursola Street
Valletta 15400
MALTA

Tel.   (356) 222645
Fax   (356) 243664

Mauritanian Red Crescent
B.P. 344
Nouakchott
MAURITANIA

Tel.   (222)(2) 51249
Fax   (222)(2) 51249
Tlx   5830 CRM

Mauritius Red Cross Society
Ste. Thérèse Street
Curepipe
MAURITIUS

Tel.   (230) 6763604
sTlxtch YBRAT IW* 4258
       "For Mauritius Red Cross"
Tlg.   MAUREDCROSS CUREPIPE

Mexican Red Cross
Calle Luis Vives 200
Colonia Los Morales Polance
Mexico, D.F. 11510
MEXICO

Tel.   (52)(5) 3951111(ext.145) /
       3950606
       5800070 (ext.106) / 5575270
Fax   (52)(5) 3951598 / 3950044
Tlx   01777617 CRMEME
Tlg.   CRUZROJA MXICO

Red Cross of Monaco
27, boulevard de Suisse
Monte Carlo
MONACO

Tel.   (33)(93) 506701
Fax   (33)(93) 159047
Tlg.   CROIXROUGE
       MONTECARLO

Red Cross Society of Mongolia
Central Post Office
Post Box 537
Ulaanbaatar
MONGOLIA

Tel.   (976)(1) 320934 /
       323334 (President's office)
Fax   (976)(1) 320934
Tlx   79358 MUZN
Tlg.   MONRECRO
       ULAANBAATAR

Moroccan Red Crescent
B.P. 189
Rabat
MOROCCO

Tel.   (212)(7) 650898 / 651495
Fax   (212)(7) 759395
Tlx   ALHILAL 319-40 M RABAT
Tlg.   ALHILAL RABAT

Mozambique Red Cross Society
Caixa Postal 2986
Maputo
MOZAMBIQUE

Tel.   (258)(1) 430045/7
Fax   (258)(1) 429545
Tlx   6-169 CV MO

Myanmar Red Cross Society
Red Cross Building
42 Strand Road
Yangon
MYANMAR

Tel.   (95)(1) 95232
Fax   (95)(1) 96551
Tlx   21218 BRCROS BM
Tlg.   MYANMARCROSS
       YANGON

Namibia Red Cross
P.O. Box 346
Windhoek
NAMIBIA

Tel.   (264)(61) 235216 / 235226 /
       235348 / 235346
Fax   (264)(61) 228949

Nepal Red Cross Society
P.O. Box 217
Kathmandu
NEPAL

Tel.   (977)(1) 270650 / 270761 /
       272761 / 270167
       278719 / 273734
Fax   (977)(1) 271915
Tlx   2569 NRCS NP
Tlg.   REDCROSS KATHMANDU
Eml   ifrc@ifrcpc.mos.com.np

The Netherlands Red Cross
Postbus 28120
2502 KC The Hague
NETHERLANDS

Tel.   (31)(70) 3846868 / 3846764
       (International Department)
Fax   (31)(70) 3846643
       (Information Department &
       General Number)
       3846751 (International
       Activities)
       3846763 (Board of
       Directors/International
Secretariat)
Tlx   32375 NRCS NL
Tlg. ROODKRUIS THE HAGUE
Eml   hq@redcross.nl

New Zealand Red Cross
P.O. Box 12140
Thorndon
Wellington
NEW ZEALAND
Tel. (64)(4) 4723750
Fax   (64)(4) 4730315

Nicaraguan Red Cross
Apartado 3279
Managua
NICARAGUA

Tel.   (505)(2) 652082 / 652084
Fax   (505)(2) 651643
Tlx   2363 NICACRUZ
Tlg. NICACRUZ-MANAGUA
Eml   nicacruz@com.ni

Red Cross Society of Niger
B.P. 11386
Niamey
NIGER

Tel.   (227) 733037 / 722706
Tlx   CRN GAP NI 5371

Nigerian Red Cross Society
P.O. Box 764
Lagos
NIGERIA

Tel.   (234)(1) 683701 / 681327 /
       680865
Tlx   21470 NCROSS NG
Tlg. NIGERCROSS LAGOS

Norwegian Red Cross
Postbox 6875
St. Olavsplass
0131 Oslo
NORWAY

Tel.   (47)(22) 943030
Fax   (47)(22) 206840
Tlx   76011 NORCR N
Tlg. NORCROSS OSLO
Eml   roajoh@redcross.no

Pakistan Red Crescent Society
Sector H-8
Islamabad
PAKISTAN

Tel.   (92)(51) 854885 / 856420
Fax   (92)(51) 280530
Tlx   54103 PRCS PK
Tlg. HILALAHMAR
       ISLAMABAD

Red Cross Society of Panama
Apartado 668
Zona 1 Panama
PANAMA

Tel.   (507) 2283014 / 2282786 /
       2280692
Fax   (507) 2286857
Tlx   2661 STORTEXPA
Tlg. PANACRUZ PANAMA

**Papua New Guinea Red Cross
Society**
P.O. Box 6545
Boroko
PAPUA NEW GUINEA

Tel.   (675) 258577 / 258759
Fax   (675) 259714
Tlx   PNG RC NE 23292

Paraguayan Red Cross
Brasil 216 esq. Jos Berges
Asunción
PARAGUAY

Tel.   (595)(21) 22797 / 208199 /
       205496
Fax   (595)(21) 211560
Tlg. CRUZ ROJA PARAGUAYA

Peruvian Red Cross
Apartado 1534
Lima
PERU

Tel.   (51)(14) 4482005. 4489431.
       4481653
Fax   (51)(14) 4486472
Tlx   21002 -cp CESAR
       25202 -cp CESAR
       "Para Cruz Roja Peruana"
Tlg. CRUZROJA PERUANA
       LIMA

**The Philippine National Red
Cross**
P.O. Box 280
Manila 2803
PHILIPPINES

Tel.   (60)(2) 5278384 to 98
       (main line)
Fax   (60)(2) 5270857 (direct fax
       line)
Tlx   27846 PNRC PH
Tlg. PHILCROSS MANILA
Eml   pnrc@phil.gn.apc.org

Polish Red Cross
P.O Box 47
00-950 Varsovie
POLAND

Tel.   (48)(2) 6285201-7
Fax   (48)(2) 6284168
Tlx   813561 PCK PL
Tlg. PECEKA VARSOVIE

Portuguese Red Cross
Jardim 9 de Abril, 1 a 5
1293 Lisboa Codex
PORTUGAL

Tel.   (351)(1) 605571 / 605650
       605490 / 3962127
       (International Department)
Fax   (351)(1) 3951045
Tlg. CRUZVERMELHA

Qatar Red Crescent Society
P.O. Box 5449
Doha
QATAR
Tel.   (974) 435111
Fax   (974) 439950
Tlx   4753 qrcs dh
Tlg. hilal doha

Romanian Red Cross
Strada Biserica Amzei, 29
Sector 1
Bucarest
ROMANIA

Tel.   (40)(1) 6593385 / 6506233
       6503813 / 3124026
       (Int. Dept)
Fax   (40)(1) 3128452
Tlx   10531 romcr r
Tlg. ROMCROIXROUGE
       BUCAREST

The Russian Red Cross Society
Tcheryomushkinski Proezd 5
117036 Moscow
RUSSIAN FEDERATION

Tel.   (7)(095) 1266770 / 1261403
Fax   (7)(095) 2302867
Tlx   411400 IKPOL SU
Tlg. IKRESTPOL MOSKWA

Rwandan Red Cross
B.P. 425
Kigali
RWANDA

Tel.   (250) 73302 / 74402 / 75088
Fax   (250) 22558 UNHCR "Pour
       C.R."
Tlx   22663 CRR RW

**Saint Kitts and Nevis Red
Cross Society**
P.O. Box 62
Basseterre
SAINT KITTS AND NEVIS

Tel.   (1)(809) 4652584
Fax   (1)(809) 4652584

Saint Lucia Red Cross
P.O. Box 271
Castries St Lucia, W.I.
SAINT LUCIA

Tel.   (1)(809) 4525582
Fax   (1)(809) 45 37811
       (45 32358 Mr. Kenneth
       Monplaisir)
Tlx   6256 MCNAMARA LC
       Attn. Mrs Boland

**Saint Vincent and the
Grenadines Red Cross**
P.O. Box 431
SAINT VINCENT AND
THE GRENADINES

Tel.   (1)(809) 4571816 / 4571381
Fax   (1)(809) 4572235 / 4561648
       (in case of malfunction)
Tlx   7538 TA VQ "For Red Cross"

**Western Samoa Red Cross
Society**
P.O. Box 1616
Apia
SAMOA

Tel.   (685) 22676
Fax   (685) 23722
Tlx   779 224 MORISHED SX
       (Attention Red Cross)

Red Cross of San Marino
Via Scialoja, Cailungo
Rep. San Marino, 47031
SAN MARINO

Tel.   (39)(549) 994360
Fax   (39)(549) 903967/994360
Tlg. CROCE ROSSA
       REPUBBLICA
       DI SAN MARINO R.S.M.

**Sao Tomé and Principe Red
Cross**
B.P. 96
Sao Tomé
SAO TOME AND PRINCIPE

Tel.   (239)(12) 22305 / 22469
Fax   (239)(12) 21365 publico ST
Tlx   213 PUBLICO ST pour
       "Croix-Rouge"

**Saudi Arabian Red Crescent
Society**
General Headquarters
Riyadh 11129
SAUDI ARABIA

Tel.   (966)(1) 4067956
Fax   (966)(1) 4042541
Tlx   400096 HILAL SJ

Senegalese Red Cross Society
B.P. 299
Dakar
SENEGAL

Tel.   (221) 233992
Fax   (221) 225369
Tlx   61206 CSR SG

Seychelles Red Cross Society
B.P. 53
SEYCHELLES

Tel.   (248) 322122
Fax   (248) 322122
Tlx   2302 HEALTH SZ

Sierra Leone Red Cross Society
P.O. Box 427
Freetown
SIERRA LEONE

Tel.   (232)(22) 222384
Fax   (232)(22) 229083
Tlx   3692 SLRCS
       3222 ALCON SL
       3523 SCB HO SL
Tlg. SIERRA RED CROSS

Singapore Red Cross Society
Red Cross House
15 Penang Lane
Singapore 0923
SINGAPORE

Tel.   (65) 3373587 / 3360269
Fax   (65) 3374360
Tlx   SRCS RS 33978
Tlg. REDCROS SINGAPORE

Slovak Red Cross
Grosslingova 24
814 46 Bratislava
SLOVAKIA

Tel.   (42)(7) 325305 (Secretary
       General - language:
       English) /
       323576 (Deputy SG -
       language: Slovak)
Fax   (42)(7) 323279 (Secretary
       General - language:
       English) / 323576
       (Deputy SG - language:
       Slovak)
Tlx   122 400 CSRC C
Tlg. CROIX PRAHA

Red Cross of Slovenia
P.O. Box 236
61111 Ljubljana
SLOVENIA

Tel.   (386)(61) 126 1200
Fax   (386)(61) 125 2142

The Solomon Islands Red Cross
P.O. Box 187
Honiara
SOLOMON ISLANDS
Tel.   (677) 22682
Fax   (677) 25299
Tlx   66347 WING HQ

Somali Red Crescent Society
c/o ICRC Box 73226
Nairobi, Kenya
SOMALIA

Tel.   Mogadishu (871
        or 873) 131 2646
        Nairobi (254)(2) 723963
Fax   1312647 (Mogadishu)
        715598 (Nairobi)
Tlx   25645 ICRC KE

The South African Red Cross Society
P.O. Box 2829
Parklands 2121
SOUTH AFRICA

Tel.   (27)(11) 4861313/4 / 6461384
Fax   (27)(11) 4861092
Tlg.   REDCROSS
        JOHANNESBURG

Spanish Red Cross
Rafael Villa, s/n (Vuelta Gins Navarro)
28023 El Plantio
Madrid
SPAIN

Tel.   (34)(1) 3354444
        3354545 (Emergencias)
Fax   (34)(1) 3354455
        3354555 (Emergencias)
Tlx   23853 OCCRE E
        47064 CRCNC (Emergencias).
Tlg.   CRUZ ROJA ESPAÑOLA
        MADRID

The Sri Lanka Red Cross Society
P.O. Box 375
Colombo
SRI LANKA

Tel.   (94)(1) 691095 / 699935
Fax   (94)(1) 695434
Tlx   23312 SLRCS CE
Tlg.   RED CROSS COLOMBO

The Sudanese Red Crescent
P.O. Box 235
Khartoum
SUDAN

Tel.   (249)(11) 72011 / 72877
Tlx   23006 LRCS SD
Tlg.   EL NADJA KHARTOUM

Suriname Red Cross
Postbus 2919
Paramaribo
SURINAME

Tel.   (597) 498410
Fax   (597) 464780
Tlx   134 SURBANK
        "Att. Mr. Linger
        for Red Cross"

Baphalali Swaziland Red Cross Society
P.O. Box 377
Mbabane
SWAZILAND

Tel.   (268) 42532
Fax   (268) 46108
Tlx   2260 WD
Tlg.   BAPHALALI MBABANE
Eml   bsrcs@wn.apc.org

Swedish Red Cross
Box 27316
S-102 54 Stockholm
SWEDEN
Tel.   (46)(8) 6655600
Fax   (46)(8) 6604586 (Office of the
        Secretary General)
        6621805 (International
        Department)
        7836692 (National Department)
        6612701 (Information
        Department)
Tlx   19613 SWECROS S
Tlg.   SWEDCROS STOCKHOLM
Eml   postmaster@redcross.se

Swiss Red Cross
Postfach
3001 Bern
SWITZERLAND

Tel.   (41)(31) 3877111
        3300222 (Lab. de transfusion
        sanguine)
Fax   (41)(31) 3877122
        3877373 (Coopération
        internationale)
Tlx   911102 CRSB CH
        32948 CR LAB
        (Lab. de transfusion sanguine)
Tlg.   CROIXROUGE SUISSE
        BERNE

Syrian Arab Red Crescent
Al Malek Aladel Street
Damascus
SYRIAN ARAB REPUBLIC

Tel.   (953)(11) 4457282
Fax   (963)(11) 4425677
Tlx   412857 HLAL
Tlg.   CROISSANROUGE DAMAS

Red Crescent Society of Tajikistan
120, ulitsa Omara Khayyama
Dushanbe
TAJIKISTAN
Tel.   (7)(3772) 240374 / 245982

Tanzania Red Cross National Society
P.O. Box 1133
Dar es Salaam
TANZANIA, UNITED
REPUBLIC OF

Tel.   (255)(51) 46855 /
        27057 (General line)
Fax   (255)(51) 476855
Tlx   TACROS 41878

The Thai Red Cross Society
Paribatra Building
Central Bureau
Rama IV Road
Bangkok-10330
THAILAND

Tel.   (66)(2) 2564037 / 2564038
Fax   (66)(2) 2553727 / 2553064
Tlx   82535 threcso th
Tlg.   THAICROSS BANGKOK
Eml   vichitra@chulkn.chula.ac.th

Togolese Red Cross
B.P. 655
Lome
TOGO
Tel. (228) 212110
Fax (228) 215228
Tlx UNDERVPRO 5261/5145
"Pour Croix-Rouge"
Tlg. CROIX-ROUGE
TOGOLAISE LOME

Tonga Red Cross Society
P.O. Box 456
Nuku'Alofa
South West Pacific
TONGA
Tel. (676) 21360 / 21670
Fax (676) 24158
Tlx 66222 CW ADM TS Attn.
Redcross
Tlg. REDCROSS TONGA

**The Trinidad and Tobago
Red Cross Society**
P.O. Box 357
Port of Spain
Trinidad
TRINIDAD AND TOBAGO
Tel. (1)(809) 6278215 / 6278128
Fax (1)(809) 6278215
Tlx (294) 9003 "For Red Cross"
Tlg. TRINREDCROSS PORT OF
SPAIN

Tunisian Red Crescent
19, Rue d'Angleterre
Tunis 1000
TUNISIA
Tel. (216)(1) 240630 / 245572
Fax (216)(1) 340151
Tlx 14524 HILAL TN
Tlg. HILALAHMAR TUNIS

Turkish Red Crescent Society
Genel Baskanligi
Karanfil Sokak No. 7
06650 Kizilay
Ankara
TURKEY
Tel. (90)(312) 4177680
Fax (90)(312) 4177682
Tlx 44593 KZLY TR
Tlg. KIZILAY ANKARA

**Red Crescent Society of
Turkmenistan**
48 A. Novoi str.
744000 Ashgabat
TURKMENISTAN
Tel. (7)(3632) 295512
Fax (7)(3632) 251750

The Uganda Red Cross Society
P.O. Box 494
Kampala
UGANDA
Tel. (256)(41) 258701. 258702
Fax (256)(41) 258184
Tlx (0988) 62118 redcrosug
Tlg. UGACROSS KAMPALA

Red Cross Society of Ukraine
30, ulitsa Pushkinskaya
252004 Kiev
UKRAINE
Tel. (7)(44) 2250157. 2250334
Fax (7)(44) 2246173
Tlx 131329 LI RED/CROSS SU

**Red Crescent Society of
the United Arab Emirates**
P.O. Box 3324
Abu Dhabi
UNITED ARAB EMIRATES
Tel. (971)(2) 661500
Fax (971)(2) 669919
Tlx 23582 RCS EM
Tlg. HILAL AHMAR ABU
DHABI

British Red Cross
9 Grosvenor Crescent
London SW1X 7EJ
UNITED KINGDOM
Tel. (44)(171) 2355454
Fax (44)(171) 2456315 (General)
2350397 (International
Division)
2355194 (Director General)
2357447 (UK Operations)
Tlx 918657 BRCS G
Tlg. REDCROS, LONDON, SW1
Eml lorna_finnegan@brcsnhq.org

American Red Cross
Office of the President
17th and D Streets, N.W.
Washington, DC 20006
UNITED STATES
Tel. (1)(202) 7286600
(1)(703) 2066156 (24-hour
Emergency
Communications)
(1)(202) 7286600/6691 (Int.
Relief and Development/
Int. Field Personnel/
Technology Transfer and
Int. Humanitarian Law)
Fax (1)(202) 7750733 (Gen.
Manager, Int. Services)
(1)(202) 7286404 (Director, Int.
Relief and Development)
(1)(703) 2066181 (24-hour
Emergency
Communications)
(1)(202) 7286437 (Int. Field
Personnel/Transfer
Technology and Int.
Humanitarian Law)
Tlx ARC TLX WSH 892636
Tlg. AMCROSS WASHINGTON
DC
Eml postmaster@usa.red-
cross.org

Uruguayan Red Cross
Avenida 8 de Octubre, 2990
11600 Montevideo
URUGUAY
Tel. (598)(2) 800714 / 802112
Fax (598)(2) 800714
Tlg. CRUZ ROJA
URUGUAYA MONTEVIDEO

**Red Crescent Society of
Uzbekistan**
30, Yusuf Hos Hojib St.
700031 Tashkent
UZBEKISTAN
Tel. (7)(3712) 563741
Fax (7)(37712) 561801

Vanuatu Red Cross Society
P.O. Box 618
Port Vila
VANUATU
Tel.   (678) 22599
Fax   (678) 22599
Tlx   VANRED
Tlg.   VANRED

Venezuelan Red Cross
Apartado 3185
Caracas 1010
VENEZUELA

Tel.   (58)(2) 5714380 / 5712143
        5715435 / 5715957
Fax   (58)(2) 5761042 / 5715435
Tlx   27237 CRURO VC
Tlg.   CRUZ ROJA CARACAS

Red Cross of Viet Nam
68, rue Ba Triu
Hanoi
VIET NAM

Tel.   (84)(4) 262315 / 264868 /
        266283/4
Fax   (84)(4) 266285
Tlx   411415 VNRC VT
Tlg.   VIETNAMCROSS HANOI
Eml   nvh@netnam.org.vn

Yemen Red Crescent Society
P.O. Box 1257
Sanaa
YEMEN

Tel.   (967)(1) 283132 / 283133
Fax   (967)(1) 283131
Tlx   3124 HILAL YE
Tlg.   SANAA HELAL AHMAR

Yugoslav Red Cross
Simina 19
11000 Beograde
YUGOSLAVIA

Tel.   (381)(11) 623564
Fax   (381)(11) 622965
Tlx   11587 YU CROSS
Tlg.   YUGOCROSS BELGRADE

Red Cross Society of the
Republic
of Zaïre
B.P. 1712
Kinshasa I
ZAIRE

Tel.   (243)(112) 31096
Tlx   21301
Tlg.   ZAIRECROIX KINSHASA
        (BP 1712)

Zambia Red Cross Society
P.O. Box 50001
(Ridgeway 15101)
Lusaka
ZAMBIA

Tel.   (260)(1) 250607 / 254798
Fax   (260)(1) 252219
Tlx   ZACROS ZA 45020
Tlg.   REDRAID LUSAKA
Eml   zrcs@zamnet.zm

Zimbabwe Red Cross Society
P.O. Box 1406
Harare
ZIMBABWE

Tel.   (263)(4) 724653/4
Fax   (263)(4) 751739
Tlx   24626 ZRCS ZW
Tlg.   ZIMCROSS HARARE

Worldwide support: Whenever and wherever international assistance flows to help a National Red Cross or Red Crescent Society with disaster relief, delegates from the global humanitarian network of the International Federation can be found, providing technical advice to the National Society, assisting it to build effective reporting systems and helping ensure that the immediate relief needs are met efficiently while finding ways to enhance and strengthen National Society capacities to deal with future disasters.

Displaced camp health check, Azerbaijan, 1994. Ian Berry/Magnum

The International Federation network

Contact details for regional and national delegations of the International Federation of Red Cross and Red Crescent Societies.

Information correct as of 1 March 1996.
Please forward any corrections to the Federation's Information Resource Centre in Geneva.

THE INTERNATIONAL FEDERATION OF RED CROSS AND RED CRESCENT SOCIETIES
P.O. Box 372
1211 Genève 19
SWITZERLAND

Tel.   (41)(22) 7304222
Fax   (41)(22) 7330395
Tlx   (045) 412133 FRC CH
Tlg.   LICROSS GENEVA
Eml   secretariat@ifrc.org

Regional Delegations

Buenos Aires
Tucuman 731 2o.R
1049 Buenos Aires
ARGENTINA

Tel.   (54)(1)3254995
Fax   (54)(1) 3228451
Eml   argentin@ifrc.satlink.net

Sydney
7th floor, Red Cross House
159 Clarence Street
Sydney NSW 2000
AUSTRALIA

Tel.   (61)(2) 2992320
Fax   (61)(2) 2992320

Brazzaville
Zone Industrielle de MPILA
Ancien immeuble O.C.V., 4ème étage
BP 88
Brazzaville
CONGO

Tel.   (242) 824031
Fax   (242) 824785

San José
Apartado 7-3320
San José 1000
COSTA RICA

Tel.   (506) 2326565 / 2327575
Fax   (506) 2328383
Eml   fedecruz@sol.rasca.co.cr

Abidjan
B.P. 2090
Abidjan 04
COTE D'IVOIRE

Tel.   (225) 212891
Fax   (225) 212562
Tlx   22673 LRCS CI
Eml   regdel@regdel.ifrc.ci

Budapest
Zolyomi Lepcso Ut 22
Arnay J.U. 31 Pf. 121
1367 Budapest 12
HUNGARY

Tel.   (36)(1) 1869627 / 1869606 /
          2093431/2093442/2093484
Fax   (36)(1) 1869275 / 1821642
          (HoD private fax)
Eml   healey@ifrc.hu

New Delhi
2-C2 Parkwood Apartments
Rao Tula Ram Marg (opp.
Shanti Niketan)
New Delhi - 110 022
INDIA

Tel.   (91)(11) 6114760 / 6114762
Fax   (91)(11) 6116661

Kingston
P.O. Box 1284
Kingston 8
JAMAICA

Tel.   (1)(809) 9268329
Fax   (1)(809) 9683497

Amman
P.O. Box 830511 / Zahran
Amman
JORDAN

Tel.   (962)(6) 681060
Fax   (962)(6) 694556
Eml   ifrcjo01@mtc.gn.apc.org

Almaty
c/o Red Cross and Red Crescent
Society of Kazakhstan
86, Ulitsa Karla Marxa
480100 Almaty
KAZAKHSTAN

Tel.   (7)(3272) 618838
Fax   (7)(3272) 541535
Tlx   (064) 212378 ifrc su
Eml   /pn=ifrckz.ala/o=customer
          /admd=kazmail/c=kz/
          @gateway.sprint.com

Nairobi
Woodlands road (off Lenana
Road)
P.O. Box 41275
Nairobi
KENYA

Tel.   (254)(2) 714256 /714313 /
          714314
Fax   (254)(2) 718415
Tlx   22622 IFRCKE
Eml   ifrcke01@ifrc.org

Kuala Lumpur
c/o Malaysian Red Crescent
Society
32, Jalan Nipah
Off Jalan Ampang
55000 Kuala Lumpur
MALAYSIA

Tel.   (69)(3) 4510723
Fax   (60)(3) 4519359
Eml   regdel@ifrc.po.my

Moscow
c/o Russian Red Cross
Tcheremushkinski Proezd 5
117036 Moscow
RUSSIAN FEDERATION

Tel.   (7)(095) 2306620 / 2306621
Fax   (7)(095) 2306622
Eml   rcmos@glas.apc.org

Harare
11, Phillips Avenue
Belgravia
Harare
ZIMBABWE

Tel.   (263)(4) 703593 / 705166 /
          705167 / 720315 / 720316
Fax   (263)(4) 708784
Tlx   24792
Eml   ifrc@mango.zw

Country Delegations

Afghanistan
c/o Asia/Pacific Department
International Federation of Red
Cross and Red Crescent Societies
P.O. Box 372
1211 Geneva 19
SWITZERLAND

Tel.   (92)(51) 823980 (Islamabad)
          (92)(521) 843116 (Peshawar)
Fax   (92)(51) 818606 (Islamabad)
          (92)(521) 843116 (Peshawar)

Albania
c/o Albanian Red Cross
RRuga "Muhamet Gjallesha"
Tirana
ALBANIA

Tel.   (355)(42) 27508 / 25855
Fax   (355)(42) 27508 / 25855

Angola
Caixa Postal 3324
Luanda
ANGOLA

Tel.   (244)(2) 3244448
Fax   (244)(2) 320648 (also phone)
Tlx   3394 crzver an
Eml   ifrclad@angonet.gn.apc.org

Armenia
Djrashati Street 96
Yerevan-19
ARMENIA

Tel.   (7)(8852) 522253 / 561889 /
          561768
Tlx   243328 rcaid su
Eml   redplus@arminco.com

Azerbaijan
Niazi Street 11
Baku 370000
AZERBAIJAN

Tel.   (99)(8412) 931889
Fax   (99)(412) 931889
Tlx   (064) 142454
Eml   root@ifrc.baku.az

Bangladesh
c/o Bangladesh Red Crescent
Society
684-686 Bara Magh Bazar
Dhaka - 1217
BANGLADESH

Tel.   (880)(2) 834701
Fax   (880)(2) 835148
Eml   hod@ifrcbd.pradeshta.net

Benin
B.P. 08-1070
Lot A9 - Les Cocotiers
Haie Vive
Cotonou
BENIN
Tel.   (229) 301013
Fax   (229) 300614

Bosnia-Herzegovina
Trumbiceva Obala 18
58000 Split
CROATIA

Tel.   (385)(21) 357035 / 357039 /
      356642 / 357040
Fax   (385)(21) 357045
Eml   ifrc@public.srce.hr

Burundi
B.P. 324
Bujumbura
BURUNDI

Tel.   (257) 229524 / 229525 /
      229528
Fax   (257) 229408

Cambodia
Central Post Office/P.O. Box 620
Phnom Penh
CAMBODIA

Tel.   (855) 2326370
Fax   (855) 2326599
Eml   ifrckh01@uni.fi

Congo
Comité de la Croix-Rouge
congolaise
B.P. 650
Pointe-Noire
CONGO

Tel.   (242) 945471
Fax   (242) 945471

Côte d'Ivoire
c/o Croix-Rouge de Côte
d'Ivoire
B.P.1244
Abidjan 01
COTE D'IVOIRE

Tel.   (225) 321529
Fax   (225) 225355

Croatia
Ivana Lucica 6/111
10000 Zagreb
CROATIA

Tel.   (385)(1) 6110155
Fax   (385)(1) 6112135
Eml   ifrc@aixesa.srce.hr

Eritrea
c/o Red Cross Society of Eritrea
P.O. 575
Asmara
ERITREA

Tel.   (291)(1) 127857
Fax   (291)(1) 124198 / 125965
      (ICRC)

Ethiopia
c/o Ethiopian Red Cross
P.O. Box 195
Addis Ababa
ETHIOPIA

Tel.   (251)(1) 514571
Fax   (251)(1) 512888
Tlx   21338 ERCS
Eml   ifrcet01@padis.gn.apc.org

Georgia
35, Tevdore Mgvdeli Street
Tbilisi
GEORGIA

Tel.   (7)(8832) 950945 / 951404 /
      351167 / 340854
Fax   (49)(51)
      5113068/86/67/60/57/59
      (satellite fax)
Tlx   (064) 212378
Eml   root@ifrc.aod.ge

Ghana
Ring Road Central House
No C 263/3
1st floor
next to Provident Insurance
Tower
Accra
GHANA

Tel.   (233)(21) 232133
Fax   (233)(21) 232133

Guinea
c/o Croix-Rouge de Guinée
Boîte postale No 376
Conakry
GUINEA

Tel.   (224) 414255
Fax   (224) 414255

Haiti
BP 19154
Port-au-Prince
HAITI

Tel.   (509) 577212
Fax   (509) 234787 (normal tariff)
      (874) 1300726
      (Inmarsat $6/min.)

India
2-C2 Parkwood Apartments
Rao Tula Ram Marg (opp.
Shanti Niketan)
New Delhi - 110 022
INDIA

Tel.   (91)(11) 6114760 / 6114762
Fax   (91)(11) 6116661
Eml   ifrc@ifrcin.unv.ernet.in

**Lao People's Democratic
Republic**
P.O. Box 2948
Vientiane
LAO PEOPLE'S DEMOCRATIC
REPUBLIC

Tel.   (856)(21) 216610
Fax   (856)(21) 215935
Tlx   4491 TE VTE LS

Lebanon
N. Dagher Building
Mar Tacla - Beirut
LEBANON

Tel.   (971)(1) 424851
         (873) 1754315 (satellite tel.)
Fax   (873) 1754316 (satellite fax)

Liberia
c/o Liberian Red Cross Society
National Headquarters
P.O. Box 5081
Monrovia
LIBERIA

Tel.   (231) 225990
Fax   (231) 225173 (ICRC)

Malawi
c/o Malawi Red Cross
P.O. Box 30096
Lilongwe 3
MALAWI

Tel.   (265) 732877 / 732878
Fax   (265) 731403
Tlx.   44276
Eml   ifrc@unima.wn.apc.org

Mali
Fédération CR/CR au Mali
c/o Croix-Rouge malienne
B.P. 290
Bamako
MALI

Tel.   (223) 226546
Fax   (223) 226456

Mongolia
c/o Red Cross Society of
Mongolia
Central Post Office
Post Box 537
Ulaan Baatar
MONGOLIA

Tel.   (976)(1) 321684
Fax   (976)(1) 321684
Tlx   79358 MUZN

Mozambique
Caixa postal 2488
Maputo
MOZAMBIQUE

Tel.   (258)(1) 421210
Fax   (258)(1) 423507
Tlx   6169 CV Mo
Eml   saenz@ifrc.uem.mz

Myanmar
c/o Myanmar Red Cross Society
Red Cross Building
42 Strand Road
Yangon
MYANMAR

Tel.   (95)(1) 95232
Fax   (95)(1) 31432
Tlx   21218

Namibia
c/o Namibia Red Cross Society
P.O. Box 346
Windhoek
NAMIBIA

Tel.   (264)(61) 222135
Fax   (264)(61) 228949

Nepal
P.O. Box 217
Kathmandu
NEPAL

Tel.   (977)(1) 273747
Fax   (977)(1) 273747
Tlx   2569 NRCS NP
Eml   ifrc@ifrcpc.mos.com.np

Papua New Guinea
c/o Regional Delegation,
Sydney
7th Floor, Red Cross House
159 Clarence Street
Sydney NSW 2000
AUSTRALIA

Fax   (61)(2) 2992320

Philippines
c/o The Philippine National Red
Cross
Bonifacio Drive, Port Area
Manila 2803
PHILIPPINES

Tel.   (63)(2) 5300049
Fax   (63)(2) 5300049
Eml   ifrc@phil.gn.apc.org

Russian Federation
P.O. Box 1019
Khabarovsk 680000
RUSSIAN FEDERATION

Tel.   (7)(4212) 334602
Tlx   (064) 141178 DVKQ

Sierra Leone
c/o Sierra Leone Red Cross
Society
P.O. Box 427
Freetown
SIERRA LEONE

Tel.   (232)(22) 222384 / 228180
Fax   (232)(22) 222384 / 228180
Tlx   3692 SLRCS SL

Somalia
P.O. Box 41275
Nairobi
KENYA

Tel.   (254)(2) 564602 /564623
Fax   (254)(2) 564750
Tlx   25436 IFRC KE

Sri Lanka
120 Park Road
Colombo 5
SRI LANKA

Tel.   (94)(1) 592159 (direct) /
         581903 (general)
Fax.   (94)(1) 583269

Sudan
P.O. Box 10697
East Khartoum
SUDAN

Tel.   (249)(11) 70484 / 71033
Fax   (249)(11) 770484
Tlx   23006 LRCS SD

Tanzania, United Republic of
c/o Tanzania Red Cross
Upanda Road
P.O. Box 1133
Dar es Salaam
TANZANIA, UNITED
REPUBLIC OF

Tel. (255)(51) 75214 / 75139 /
      75148
Fax (255)(51) 75214 / 75139 /
      75148
Tlx 41669 TANDAR

Turkey
Ataturk Bulvari
219/14 Bulvari Apt.
006680 Kavaklidere
Ankara
TURKEY

Tel. (90)(312) 4672099 / 4673349
      4674383 / 4674384
Fax (90)(312) 4274217
Eml ifrc-o@servis2.net.tr

Uganda
c/o Uganda Red Cross Society
P.O. Box 494
Kampala
UGANDA

Tel. (256)(41) 231480 / 243742
Fax (256)(41) 258184
Eml ifrcug@imul.com

Ukraine
c/o Red Cross Society of
Ukraine
30, Ulitsa Pushkinskaya
252004 Kiev 4
UKRAINE

Tel. (7)(44) 2286110
Fax (7)(44) 2245082
Tlx (064) 131329

Viet Nam
c/o UNDP Viet Nam
P.O. Box 618
Bangkok 10501
THAILAND

Tel. (84)(4) 252250 / 229283
Fax (84)(4) 266177
Tlx 411415 VNRC VT
Eml ifrcvn@netnam.org.vn

Yugoslavia
Simina Ulica Broj 21
11000 Belgrade
YUGOSLAVIA

Tel. (381)(11) 3282202 / 3281376 /
      3282253
      3281582 / (873)
      1754574 (satellite tel.)
Fax (381)(11) 3281791
Eml ifrcbgd@telekom.etf.bg.ac.yu

Zaire
Concession Boukin, 4
Z/Ngaliema, Kin 1
Kinshasa
ZAIRE

Tel. (243)(88) 40182

Zambia
c/o Zambia Red Cross Society
P.O. Box 50001
Lusaka
ZAMBIA

Tel. (260)(1) 250607 / 254798
Fax (260)(1) 252219
Tlx 45020 ZACROSS ZA

International Federation disaster response around the world

The International Federation of Red Cross and Red Crescent Societies responds to the needs of disaster victims regardless of the cause of those needs or the country in which they are. Its prime mode of response is via the 169 member National Societies. They regularly mount relief work for disasters and crises in their own countries. On occasions the scale of the disaster may overwhelm the resources of the National Society, which appeals to the rest of the International Federation for assistance. This international disaster response is orchestrated by the Federation's Secretariat in Geneva.

In 1995 appeals totalling 327.4 million Swiss francs (Sfr.) were made in this way to assist some 13 million beneficiaries. As has become the norm in recent years, the single largest category of people requiring assistance is those fleeing their homes, whether as internally displaced or refugees. Of increasing significance is the growing amount of assistance going to people caught up in the social and economic dislocation of so many countries undergoing rapid change.

When looked at regionally, international assistance to disasters in Africa and Europe dominates the Federation's response. Both continents contain many countries undergoing rapid change; the Federation responds to the impact of such change and the associated stresses.

Figure 15.1: Appeals by type of disaster in 1995 (in millions Sfr.)

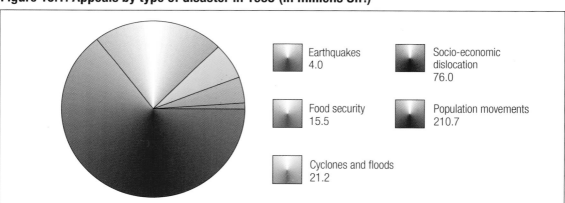

Earthquakes
4.0

Socio-economic dislocation
76.0

Food security
15.5

Population movements
210.7

Cyclones and floods
21.2

Figure 15.2: Appeals by region in 1995 (in millions Sfr.)

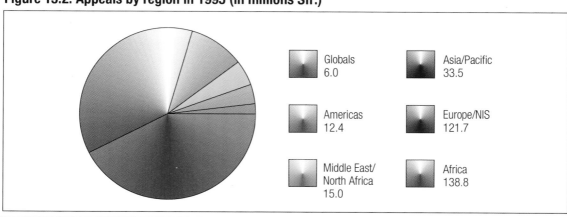

Globals
6.0

Asia/Pacific
33.5

Americas
12.4

Europe/NIS
121.7

Middle East/
North Africa
15.0

Africa
138.8

Figure 15.3: Appeals by type, number of victims and value in 1995 in date order

Country/programme	Type	No. of victims to be assisted	Appeal value (Sfr)
Eritrea	Population movement	30,000	1,176,000
Ethiopia	Food security	95,000	9,537,000
Kenya	Population movement	100,000	5,520,000
Rwanda refugees	Population movement	954,000	85,150,000
Somalia	Population movement	300,000	2,584,000
Sudan	Population movement	424,000	2,834,000
Uganda	Population movement	120,000	5,020,000
East Africa	Regional programmes	—	2,340,000
Benin/Ghana/Togo	Population movement	25,000	873,000
Liberia and region	Population movement	1,334,000	4,312,000
West Africa	Regional programmes	—	1,070,000
Central Africa	Regional programmes	—	553,000
Angola	Population movement	150,000	7,002,000
Congo/Zaire	Population movement	30,000	792,000
Malawi	Population movement	100,000	—
Mozambique	Population movement	150,000	2,652,000
Southern Africa	Regional programmes	—	1,554,000
Haiti	Socio-economic dislocation	100,000	3,989,000
Caribbean	Regional programmes	—	686,000
Central/South America	Regional programmes	—	1,238,000
South. Cone South America	Regional programmes	—	461,000
Afghanistan	Socio-economic dislocation	820,000	4,112,000
Bangladesh Myanmar refugees	Population movement	100,000	460,000
Cambodia	Socio-economic dislocation	550,000	3,432,000
Nepal	Population movement	89,000	1,020,000
Sri Lanka	Population movement	47,000	858,000
Vietnamese boat people	Population movement	13,000	493,000
South Asia	Regional programmes	—	292,000
East Asia	Regional programmes	—	1,357,000
Pacific	Regional programmes	—	1,565,000
Former Yugoslavia	Population movement	835,000	38,862,000
East and Central Europe	Regional programmes	—	1,482,000
Belarus/Moldova/Ukraine	Socio-economic dislocation	875,000	4,858,000
Chernobyl	Nuclear disaster	120,000	1,463,000
Caucasus	Population movement	655,000	25,726,000
Central Asia	Socio-economic dislocation	1,200,000	8,803,000
Russian Federation	Socio-economic dislocation	485,000	4,294,000
Iraq/Jordan/Lebanon/Palestine RC/Syria	Regional programmes	—	852,000
Iraq	Socio-economic dislocation	300,000	14,110,000
Disaster Relief Emergency Fund	—	—	6,000,000
Zaire	Epidemics		664,000
Bangladesh	Floods	100,000	1,202,000
Southern Africa	Food security	300,000	3,790,000
Eastern Europe	Epidemics	—	10,891,000
China	Floods	2,000,000	10,256,000
Cuba	Floods	5,000	570,000
Krajina	Population movement	175,000	19,680,000
Croatia	Population movement	50,000	5,657,000
Caribbean islands	Cyclone	3,000	1,420,000
Cambodia	Food security	71,000	2,142,000
DPR Korea	Floods	130,000	4,981,000
Philippines	Floods	15,000	645,000
Bangladesh	Floods	100,000	733,000
Central Africa	Floods	73,800	1,369,000
Mexico	Floods/earthquakes	96,325	4,023,000
Totals		**13,120,125**	**327,405,000**

The International Federation on the Internet

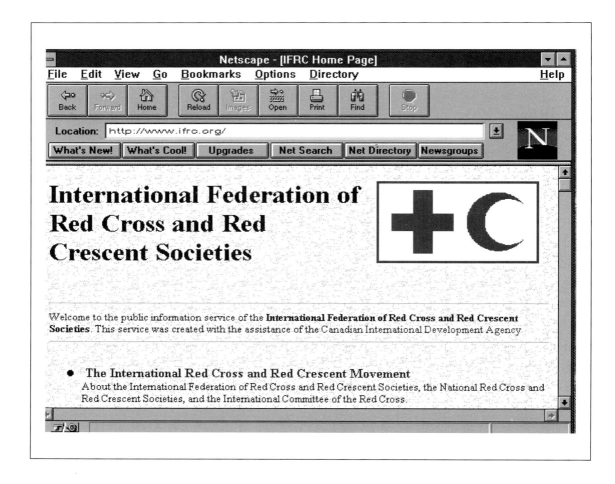

Internet users can access a wide range of information, including the full text of the *World Disasters Report*, from the International Federation of Red Cross and Red Crescent Societies.

The International Federation pages on the World-Wide Web include constantly updated disaster appeals and situation reports from its global network of operations and delegations, as well as links to many disaster-related sites and the Web pages of the following:

International Committee of the Red Cross, International Red Cross and Red Crescent Museum, American Red Cross, Australian Red Cross, British Red Cross, Canadian Red Cross, Croatian Red Cross, Danish Red Cross, Japanese Red Cross, Netherlands Red Cross, Norwegian Red Cross, Spanish Red Cross.

To make copies of publications on the Federation's Web site or to publish extracts of them, contact the Secretariat in advance for permission. Full acknowledgement will be required.

The Federation pages are at http://www.ifrc.org

If you have any problems connecting or would like more information, please contact the Federation's Internet manager, Jeremy Mortimer, on webmaster@ifrc.org

The world of disasters in the

World Disasters Report 1993-1996

The *World Disasters Report* is the only annual, global report focusing on disasters, from earthquakes to epidemics, conflict to economic crisis, and on the millions of people affected by them.

Published by the International Federation of Red Cross and Red Crescent Societies in English, French, Spanish, Japanese and now Finnish, Swedish and German, the *World Disasters Report* analyses cutting-edge issues, assesses practical methodologies, examines recent experience and collates a comprehensive disasters database.

To order the *World Disasters Report* in any of its languages and from any year, see the order form on page 174, or contact the International Federation.

The five-section *World Disasters Report 1996* includes: global population movements, causes and consequences; global food security, including issues of gender, conflict and rights; emergency food aid and nutrition; developmental relief; trends in aid; surviving the Kobe earthquake; challenges of Rwanda; Oklahoma's trauma; DPR Korea's flood and food crisis; meeting the need for systematic data; Code of Conduct update; full listings of National Societies and delegations; 25-year disasters database. Fully illustrated.

The four-section *World Disasters Report 1995* includes: UN sanctions and the humanitarian crisis, with case histories from Iraq, Serbia, Haiti; good disaster relief practice; turning early warning into livelihood monitoring; measuring the effects of evaluation; listening to the beneficiaries; psychological support; humanitarians in uniform; hate ratio in conflict; mines and demobilisation; surviving cyclones in Bangladesh; Ethiopia ten years on; success and failure in Rwanda; working in Somalia's grey zone.

The three-section *World Disasters Report 1994* includes: drought success in Southern Africa; conflict and progress in Somalia; agency challenges within the former Yugoslavia; Brazil's vulnerability; India's earthquake myths; Caucasus collapse; secrecy's role in disasters; global survey of anti-personnel mines, information and Chernobyl; African peace mechanisms; human rights and disasters; indigenous knowledge and response; and the full text of the Code of Conduct for disaster-relief agencies.

The *World Disasters Report 1993* - the pilot issue - includes: humanitarian gap, preparedness versus relief, role of foreign medical teams and military forces, equity in impact, media in disasters, AIDS, famine, flood, high winds, refugees, epidemics, earthquakes, volcanoes. Case histories from: Uganda, Sudan, China, Bangladesh, Afghanistan, Peru, Zambia, Turkey, United States, Philippines. Fully illustrated.

The *World Disasters Report Special Focus; Under the Volcanoes: Rwanda's Refugee Crisis*, is the first in an occasional series of studies covering issues of concern to policymakers, whose main conclusions will later appear in the annual *World Disasters Report. Under the Volcanoes* covers three dilemmas of Rwanda and other "total" disasters - political action versus humanitarian action, human rights versus humanitarianism, beneficiaries versus agencies - as well as a chronology of the crisis, and issues of drugs and disease, military intervention, and five ways to reduce camp violence. To order, contact the Federation.

And coming up ...

World Disasters Report 1997

In a totally new edition, the *World Disasters Report 1997* will examine the rapidly changing roles and responsibilities of governments, their peoples and aid agencies in disaster response, asking: "Who are the stakeholders in today's disasters?" With its new five-section format, the *World Disasters Report 1997* will analyse key issues, highlight vital methodologies, consider the year of disasters 1996, include its comprehensive 25-year disasters database, and outline the global activities of the International Red Cross and Red Crescent Movement. Plus: full references to the world's most useful disaster-related Internet Web sites. For further information, contact the Federation.

WORLD DISASTERS REPORT 1996 – ORDER FORM

To order the *World Disasters Report 1996* in English, please contact Oxford University Press:

> Call the 24-hour global hotline (44) (1536) 454534.

> Order through any bookshop worldwide: quote title, publisher and ISBN (0-19-829079-9 softback, 0-19-829080-2 hardback).

> Send a copy of the form below to the 24-hour global faxline (44) (1536) 454518.

> Write using a copy of this form to Andrea Nicholls, Oxford University Press, Walton Street, Oxford OX2 6DP, UK

Prices: softback GB £15.99, US $29.95 (approximately Sfr. 30.00); hardback GB £35.00, US $65.00 (approximately Sfr. 70.00). Plus post and packing costs for mail order.

Please send me _____ copy/copies of the *World Disasters Report 1996*. Total order value £/$ _____ plus post and packing.

Payment by credit card: Amex: ☐   MC/EC: ☐   Visa: ☐

(For other forms of payment and ordering, contact OUP.)

Name of card holder: ....

Card number: ....   Expiry date: ....

Signature: ....   Date: ....

Delivery to:

Name: ....

Address: ....

Post/zip code: ....   Country: ....

Telephone: ....   Fax: ....   Email: ....

Organisation: ....

Areas of interest: ....

World Disasters Report – other languages, other years

The *World Disasters Report* has been published annually since 1993 in variety of languages, including Arabic, English, French, Finnish, German, Japanese, Spanish and Swedish. Please contact the Federation for details of availability, cost and ordering. World Disasters Report, International Federation of Red Cross and Red Crescent Societies, PO Box 372, 1211 Geneva 19, Switzerland; fax (41) (22) 733 0395; tel (41) (22) 730 4222; email walker@ifrc.org

Index

A

Afghanistan, 14, 19, 24, 35, 50, 52, 79
Africa, 7, 21, 22, 49
  East, 50
  Economic Community of West
    African States, 60
  Great Lakes region, 77, 78
  Horn of, 25
  Office of Emergency Operations, 14
  sub-Saharan, 21, 24, 25, 29, 35
Anderson, Per Pinstrup, 29
Angola, 24, 36, 77
anti-personnel mines — see mines
Argentina, 31
  Red Cross Society, 31
Aristide, President Jean-Baptiste, 17
Armenia, 17
  Red Cross Society, 17
Asia, 21, 24, 26, 33, 50, 58
  Association of South East Asian
    Nations, 16
Australia, 16, 31
Austria, 31

B

Balkans, 52, 103
Bangladesh, 13, 21, 50
BBC World Service, 110
Belgium, 12
  Red Cross Society, 152
Beriberi, 37
Berlin Wall, 17
Bosnia, 15, 18, 49, 60, 63, 77, 89, 99,
  100-101, 105
  Bieljina, 106
  Bosnaplod, 49
  Brod, 101
  Derventa, 101
  Kondusa, 102
  Martin Brod, 105
  Mostar, 103
  Petrovac, 102, 106
  Prijedor, 102
  Presnace, 104
  Red Cross Society, 106
  Radinovici, 105
  Sanski Most, 102
  Sarajevo, 101
  Sava, river, 101, 104
  Srebrenica, 100
  Tuzla, 49, 103, 105

  Batva district, 105
  Centre for Social Work, 105
  Red Cross chapter, 105
  Udbina, 104
  Vrhovine, 104
  Vukosalvje, 101
  Zenica, 103
  Zepa, 100
Botswana, 28
British Red Cross Society, 91
Brown, Lester, 29
Brown University,
  World Hunger Program, 21, 26, 28
building capacity, 143
Burundi, 16, 77, 78, 82
  Kanganiro camp, 41

C

Cambodia, 12, 24
  mine-clearance efforts, 103
  human development programmes
    in, (CARERE), 12
Canada, 16, 31
  Red Cross Society, 16
CARE, 145
Caritas Internationalis, 145
Cartagena Declaration on Refugees,
  1984, 10
Catholic Church, 103
  Catholic Relief Service, 145
  Catholic Action (liberation
    theology), 23
Caucasus, 12, 25, 50
Central America, 12, 25, 78
  human development programmes
    in, (PRODERE), 12
Central Asian Republics, 12
Chechnya, 54
children, 27, 142
  World Summit for Children, 1990,
    10, 27
China, 24, 25, 27, 32, 50
  Academy of Social Sciences, 13
  government, 13
cholera, 56, 81
Cockburn, Alexander, 100
Code of Conduct, 7, 38, 39, 51, 141,
  143, 145-149
  register, 145
    Principal Commitments, 146
Cold War, 6, 55, 56, 61

complex humanitarian emergencies, 14
Comprehensive Plan of Action, 1989, (CPA), 16, 17
Coping With Crisis, 91
Cornell University, 27
Critical Incident Stress Debriefing Team, 90
Croatia, 99, 101
  Banija Region, 99
  Banja Luka, 99, 100, 107
  Benkovac, 104
  Brotherhood and Unity Highway, 105
  Donji Lapac, 107
  Glina, river, 101
  Karisnica, river, 104
  Knin, 104, 105
  Krajina, 51, 99, 100, 105, 107
  Licki Osik, 104
  Martin Brod, 106
  Medak, 104
  Obrovac, 106
  Office for Displaced Persons and Refugees, 103
  Petrovac, 107
  Red Cross Society, 99, 104, 105, 106
  Sisak, 99
  Vrhovine, 104
  Zadar, 104
  Zagreb, 101, 105
Cuba, 17, 79
  boat people, 14

D

Dayton, Peace Accords, 100
Denmark
  Copenhagen, 50
  Red Cross Society, 91, 103
    First Consultation on Psychological Support in Copenhagen, 1991, 91
    Second Consultation on Psychological Support in Copenhagen, 1995, 91
development relief, 50, 143
disaster-aid diplomacy, 114
dried, skimmed milk, 49

E

earthquakes
  Great Hanshin-Awaji, Kobe, Japan, 1995, 65
  Iran, 1991, 116
Eastern bloc, 58
Economic Community of West African States, 60
Ehrlich, Paul, 29
El Salvador, 13
emergency food rations, 27

Eritrea, 24, 26
Ethiopia, 12, 26, 42
  Office of Emergency Operations in Africa, 14
  Government Relief and Rehabilitation Administration, 116
Europe, 17, 50
  European Community, 12,31
  European Union (EU), 49
    Council of Ministers, 60
  Conference on Security and Co-operation, 17

F

families, reunification of, 143
famine and war, 142
food, 21-33
food aid, 6
  Food Aid Convention, 1986, 31
food wars, 26
France, 16
  Red Cross Society, 155

G

General Assembly of the Red Cross, 91
Geneva Conventions, 142
  Additional Protocols, 142
Germany, 58
  German request to UNHCR, 17
  Red Cross Society, 155
Ghana, 36
  Ghanaians from Nigeria, 17
global figures: refugees and IDPs, 131-135
Guatemala, 24
Gulf States, 17
Gulf war, 14, 17

H

Haiti, 14
  boat people, 14, 17
  US military occupation of, 17
Handicap International, 103
Hayashi, Professor Haruo, 73
hidden hunger, 27
hunger, 21-33
Hutus, 16, 76-87

I

Ikawa, Kasuo, 70
independence of humanitarian action, 143
India, 13
internally displaced people (IDP), 9-19, 135, 143

International Committee of the Red Cross (ICRC), 14, 18, 100, 104, 141
International Conference on Nutrition, 1992, 24
International Conference, Red Cross and Red Crescent Movement, 6, 141
  Fourth Protocol on Blinding Laser Weapons, 143
International Federation of Red Cross and Red Crescent Societies, 6, 7, 18, 52, 100, 103, 106, 111, 141
  Mobile Red Cross, 105
  Reference Centre for Psychological Support, 91
  Working Group for Psychological Support, 91
International Food Policy Research Institute (IFPRI), 29, 30, 33
  2020 Vision for Food, Agriculture and the Environment, 33
International Health Exchange, 58
International Humanitarian Law, 141-143
International Monetary Fund, 55
International Organization for Migration (IOM), 16, 17
International Red Cross and Red Crescent Movement, 7, 9, 31, 60, 104, 141
international relief system, 59
International Save the Children Alliance, 145
iodine deficiency, 21, 27, 37
Iran, 116
  earthquake, 116
  Red Crescent Society, 116
Iraq, 17, 56
  army, 14,
  Kurdish emergency, 14
  Kurds, 14
iron deficiency anaemia, 27, 37

J

Japan, 31, 58, 64-75
  Ashiy, 74
  Fukui, 66
  Himeji, 75
  Hokkaido, 68
  Hyogo, 75
    Prefecture Administration, 71, 74, 75
    Red Cross Blood Centre, 75
    Red Cross chapter, 75
  Disaster Prevention Administration (DPA), 71
  National Land Ministry, 71
  Prime Minister's Office, 71
  Red Cross Society, 74-75
  Kobe City, 65-75
    Red Cross chapter, 66
    Red Cross Hospital, 71, 74
    University, 68

Kyoto University, 73
Okayama, 74, 75
Osaka, 67, 73, 75
Otsu, 75
Tatasuki, 75
Tokyo, 66, 67, 73
    Science and Technology Agency,
    75
Totonia, 66

K

Kenya, 24, 53, 62
    Mombasa, 53
    Red Cross Society, 53
Korea, DPR, 109-117
    Amnok, river, 109
    Central Flood Damage
    Rehabilitation Committee, 111
    Huichon City, 111
    Pyongyang, 116
        Power Station, 110
    juche — self-reliance, 113
    Red Cross Society, 111-114
    Sangwol, reservoir, 109
    Sinuiju, 109, 111
Kuwait, 17
kwashiorkor, 40

L

land-mines — see mines
laser weapons, 143
    Fourth Protocol on Blinding Laser
    Weapons, 143
Latin America, 10, 21
Liberia, 42, 50, 55, 78
linking relief to development (LRD),
    25
Lockerbie, Scotland, 93
Los Angeles Times, 100
Lutheran World Federation, 145

M

Malawi, 42, 50, 62
Marxism, 23
Maya, 23
McCalla, Alex, 30
Médecins sans Frontières, 114, 145
Mexico, 23, 24
    Chiapas, 23
micronutrient deficiency problems,
    27, 31
mines
    anti-personnel, 24, 143
    campaign, 103
    UN Mine Action Centre, Zagreb, 103
    World Health Organization, 103
Mozambique, 12, 19, 25, 42, 50, 77
Murosaki, Professor, 68, 69
Myanmar, 24

N

Namibia, 12
National Red Cross and Red
    Crescent Societies, 151-154
North Atlantic Treaty Organization
    (NATO), 60, 100, 101
Norway, 31
    Norwegian Church Aid (NCA), 49
nutrition, 35-45

O

Odgovor, refugee publication, 52
oedema, 40
Oklahoma, USA, 89-97
    American Red Cross chapter, 90
    Federal Building, Alfred P. Murrah,
    89
    Funeral Directors Association, 90
    Office of the Medical Examiner, 90
    "Project Heartland", 96
    State Department of Mental Health,
    96
    National Guard, 90
    Veteran's Administration Hospital,
    90, 96
Operation Lifeline Sudan, 14
Organisation of African Unity
    (OAU), 12
    Convention Governing the Specific
    Aspects of Refugee Problems in
    Africa, 10
Organisation for Economic
    Co-operation and Development
    (OECD), 61
Oxfam, 48, 81, 154

P

Page, Trevor, 112
Pakistan, 13
PanAm, 93
Pandya-Lorch, Rajul, 29
pellagra, 35, 37, 42
post-conflict rehabilitation, 58
"Project Heartland", Oklahoma
    State, USA, 96
protection of civilians in conflict, 142

R

refugees, 8-19
    Convention relating to the Status of
    Refugees 1951, 10
    environmental, 13
    Protocol relating to the Status of
    Refugees, 1967, 10
    Refugee Nutritional Information
    System (RNIS), 39, 41
    Statistics, 122
Register of Engineers, 58

Russia, 12, 17, 25
Rwanda 12, 13, 15, 16, 24, 28, 39, 49,
    50, 55, 56, 57, 60, 63, 76-87, 89, 145
    International Tribunal, 82
    Joint Evaluation of Emergency
    Assistance, 78, 80-81
    Kigali, 81, 145

S

sanctions, 143
Save the Children Fund, UK, 83
Scandanavian Star, ferry disaster, 91
scurvy, 35, 37
Sen, Amartya, 27
Shiite Muslims, 14
Sierra Leone, 42
Slavonia, Western, 100
Somalia, 13, 15, 16, 24, 28, 53, 55, 77,
    79
South America, 24
Soviet Union, former, 17, 24, 25, 113
Sri Lanka, 13, 17, 24
Steering Committee for
    Humanitarian Response, 145
Sudan, 13, 48
    Operation Lifeline, 14
    Sudanese from Libya, 17
Sweden, 31, 59
Switzerland, 31

T

Tajikistan, 17
Tamils, 17
Tanzania, 16, 79-82
    Benaco camp, 82
    Lukole camp, 82
    Ngara, 82
    Red Cross Society, 82
television, 56
terrorism, 89
traditional birth attendant (TBA), 82
trauma, 88-97
Turkey, 14
Turkish Government, 14
Tutsis, 16, 76-87

U

Uganda, 48
United Kingdom , 61
    British Red Cross Society, 58
    Overseas Development
    Administration, 58
United Nations, 12, 18, 21, 47, 55, 60,
    61
    Centre for Human Rights, 82
    Children's Fund (UNICEF), 27, 82
    Department of Humanitarian Affairs
    (UNDHA), 15, 112

Development Programme (UNDP)
    12, 57, 80
Environment Programme
    (UNEP), 13
Food and Agriculture Organization
    (FAO), 23, 28, 110, 111
General Assembly, 14, 15
High Commissioner for Refugees
    (UNHCR), 9, 10, 11, 12, 13, 16,
    17, 18, 57, 59, 81, 103
    Special Programmes, 16
Human Rights Commissioner, 82
Mine Action Centre, Zagreb, 103
Refugee Nutrition Information
    System, 39, 41
safe havens, 100
Security Council, 14, 15, 60
    Resolution 688, 14
Subcommittee on Nutrition, 21, 35, 39
United States of America, 14, 15, 16,
    17, 21, 31, 41, 49, 59, 112
    American Red Cross Society, 90, 96,
    97
    Agency for International
    Development (USAID), 58, 59, 62

Office of Foreign Disasters
    Assistance (OFDA), 62
    Centers for Disease Control, 39, 40
    Committee for Refugees, 14, 19
    Department of Agriculture, 32
    Veteran's Administration, 90
Uvin, Professor Peter, 33

V

Viet Nam, 15, 16, 17, 50, 78, 79, 115
    Red Cross Society, 16, 115
vitamin A deficiency, 21, 27, 37

W

wasting, 36, 39, 40, 41
water and war, 143
women, 27, 28
World Bank, 13, 29, 30, 33, 55
World Council of Churches, 145
World Disasters Report 1995, 28, 77,
    87, 147
    Special Focus on Rwanda, 77, 87

World Food Programme, 9, 31, 32,
    41, 59, 110, 112
World Health Organization, 103
World Summit for Children, 27
World Trade Center bombing, 93
World War II, 55, 105
Worldwatch Institute, 22, 29, 30

Y

Young Men's Christian Association
    (YMCA), 70
Yugoslavia, former, 11, 12, 16, 24,
    55, 58, 98-107
    Federal Republic of, 100, 101, 106
Yugoslav Red Cross Society, 106-107

Z

Zaire, 12, 13, 16, 56, 79, 82
    Goma, 36, 38, 56, 63, 80
Zapatista revolution, 23, 33
Zagreb-to-Belgrade motorway, 105
Zimbabwe, 28

The State of the World's Refugees 1995
In Search of Solutions

United Nations High Commissioner for Refugees

* **Tackles an issue of huge public concern**

* **Includes detailed charts, graphs, and maps describing the state of the world's refugees**

Bosnia, Rwanda, Iraq, Somalia, Chechnya. During the past few years, the world has witnessed a succession of massive refugee movements and humanitarian emergencies.

What can be done to resolve the global refugee problem? The United Nations High Commissioner for Refugees, the international organization responsible for the world's displaced people, here examines the roots of the current crisis, and assesses the continuing relevance of traditional approaches to the problem of human displacement. While the right of asylum must be scrupulously maintained, the UNHCR suggests that greater efforts must also be made to tackle refugee problems at their source, by restoring peace and security to countries where large numbers of people have been forced to abandon their homes. To achieve this objective, concerted international action will be required to protect human rights, establish effective peacekeeping operations, and promote sustainable development.

Advisory Committee: Christoph Bertram, Robin Cohen, Dhram Ghai, Gil Loescher, Theodor Meron, Astri Suhrke, Valery Tishkov, Myron Weiner, Thomas Weiss

264 pp, numerous halftones, figures and tables, 1995
0-19-828044-0, £30.00, Hardback
0-19-828043-2, £9.99, Paperback (Oxford Paperbacks)

Green Globe Yearbook 1996
Yearbook of International Cooperation on Environment and Development

Edited by **Helge Ole Bergesen**, *Senior Research Fellow, Fridtjof Nansen Institute, and* **Georg Parmann**, *Science Editor, Scandinavian University Press*

Fifth edition of an annual publication on the international politics of environmental management consisting of 8 articles written by independent experts, focusing on the effectiveness of international environmental collaboration. A reference section of key data presents the most important international agreements on the environment and development, and inter- and non-governmental organizations that are active in this area.

'An invaluable source of information for all concerned with the complexities surrounding human use and management of the global environment ... This book will ... be of widespread benefit ... rated as excellent.' **Marine Pollution Bulletin**

376 pp, maps, March 1996
0-19-823345-0, £35.00, Hardback

MORE INFORMATION?
Contact: Andrea Nicholls, Marketing Department,
Oxford University Press, Walton Street, Oxford, OX2 6DP
Tel: +44 (0) 1865 56767 Fax: +44 (0) 1865 56646 Email: nicholla@oup.co.uk

OXFORD

Community Development Journal

Published four times a year the **Community Development Journal** covers political, economic and social programmes that link the activities of people with institutions and government. Articles feature community action, village, town and regional planning, community studies and rural development.

Volume 31, 1996 (4 issues) Institutions: £50/US$90
Individuals: £36/US$69, Developing Countries: US$65

Journal of African Economies

In the last few years there has been a growing output of high quality economic research on Africa, but until the advent of the **Journal of African Economies** it was scattered over many diverse publications. Now this important area of research has its own vehicle to carry rigorous economic analysis, focused entirely on Africa, for Africans and anyone interested in the continent - be they consultants, policymakers, academics, traders, financiers, development agents or aid workers.

Volume 5, 1996 (3 issues)
Institutions: £60/US$110, Individuals: £38/US$73
Subscribers in Africa: Institutions: US$58, Individuals: US$40

International Journal of Refugee Law

A journal which aims to stimulate research and thinking on refugee law and its development, taking account of the broadest range of State and international organisation practice. It serves as an essential tool for all engaged in the protection of refugees and finding solutions to their problems, providing key information and commentary on today's critical issues.

Volume 8, 1996 (4 issues)
Institutions: £72/US$135, Individuals: £35/US$66
Developing Countries: US$62

African Affairs

African Affairs is one of the oldest journals in the field. It publishes articles on recent political, social and economic developments in sub-Saharan countries, and includes historical studies that illuminate current events in the continent.

Volume 95, 1996 (4 issues)
Institutions: £63/US$118, Individuals: £36/US$60, Individuals in Africa: US$28

JOURNALS

Journal of Refugee Studies

The **Journal of Refugee Studies** provides a major focus for
research into refugees reflecting the diverse range of issues
involved. It aims to promote the theoretical development
of refugee studies, and encourages the voice of refugees
to be represented by analysis of their experiences.
Volume 9, 1996 (4 issues)
Institutions: £64/US$118, Individuals: £32/US$58
Institutions in Developing Countries: US$55

Refugee Survey Quarterly

NEW! RSQ is produced by the Centre for Documentation and Research of the
United Nations High Commissioner for Refugees. Published four times a year
it serves as an authoritative source for current refugee and country information.
Each issue is a combination of country reports, documents, reviews and
abstracts of refugee-related literature.
Volume 15, 1996 (4 issues)
Institutions: £55/US$85, Individuals: £38/US$60

International Journal of Epidemiology

Exploring the epidemiology of both infectious and non-infectious disease,
including research into health services and medical care, *IJE* encourages
communication among those engaged in the research, teaching, and
application of epidemiology throughout the world.
Volume 25, 1996 (6 issues)
£170/US$315

Health Policy and Planning

Particularly relevant to those working in international health planning, medical
care and public health, *Health Policy and Planning* is concerned with issues of
health policy, planning, and management and evaluation, focusing on the
developing world.
Volume 11, 1996 (4 issues)
Institutions: £100/US$180, Individuals: £43/US$80

Frank Cass

Medicine, Conflict and Survival
International Medical Concerns on Global Security Issues
(formerly Medicine and War)
Editor **Douglas Holdstock** FRCP, UK
Associate Editor **Nevin Hughes-Jones** FRCP, UK

Medicine, Conflict and Survival is an international journal for medical and health professionals and for peace researchers on: the cultural and biological causes of war and group violence; the medical and environmental effects of war and preparations for war; the ethical responsibility of health professionals in relation to war, other social violence and human rights abuses; non-violent methods of conflict resolution; medical and humanitarian aid in conflict situations; and the interaction of environmental issues with global security and the risk of armed conflict.

Recent Articles
Politics by Genocide by *P Hall* and *A Carney*
Violence in the Americas: The Emergence of a Social Epidemic by *C Chelala and G de Roux*
Health Impacts of Climate Change by *Sari Kovats and Andrew Haines*
Medicine, Population and War by *Jack Parsons*
Beyond the Line of Fire: Health in Nicaragua by *P Smith*
The Resumption of French Nuclear Testing by *K Suter*
Special Issue
Hiroshima and Nagasaki edited by *Frank Barnaby* and *Douglas Holdstock*
Forthcoming Articles
War Wounded Refugees by *Ann-Charlotte Hermansson et al*
Blinding Laser Weapons by *Ann Peters*
Living Conditions, Inequalities in Health, Human Rights and Security by *John Middleton*

ISSN 1362-3699 Volume 12 1996
Quarterly: January, April, July, October
Individuals £30/$45 Institutions £85/$120

UK/OVERSEAS ORDERS to: Frank Cass, 890–900 Eastern Avenue,
Ilford, Essex IG2 7HH, UK. Tel: 0181 599 8866
Fax: 0181 599 0984 E-mail: sales@frankcass.com
US ORDERS to: Frank Cass, c/o ISBS,
5804 N E Hassalo Street, Portland, OR 97213-3644, USA.
Tel: (503) 287-3093, (800) 944-6190 Fax: (503) 280-8832, E-mail:orders@isbs.com

Frank Cass

Environmental Politics

Editors
Stephen Young, *University of Manchester, UK*
Michael Waller, *Keele University, UK*

Environmental Politics is concerned with three particular aspects of the study of environmental politics, with a focus on industrialised countries. First, it examines the evolution of environmental movements and parties. Second, it provides an analysis of the making and implementation of public policy in the area of the environment at international, national and local levels. Third, it carries comment on ideas from both a 'deep' and a 'shallow' perspective, generated by the various environmental movements and organisations, and by individual theorists. It is sensitive to the distinction between goals of conservation and of a radical reordering of political and social preferences, and aims to explore the interface between these goals, rather than to favour any one position in contemporary debates. Each issue contains a number of full-length articles; a profile section; and a bibliographical section. The profile section in each number focuses on current issues, providing a first perspective and analysis. It is of particular value for teaching, or simply for keeping abreast of important developments in the politics of the environment.

Recent Articles

Britain and the Intergovernmental Panel on Climate Change: The Impacts of Scientific Advice on Global Warming Part 1: Integrated Policy Analysis and the Global Dimension by *S A Boehmer-Christiansen*
The Global Environment Facility in its North-South Context by *Joyeeta Gupta*
Britain and the Intergovernmental Panel on Climate Change: The Impacts of Scientific Advice on Global Warming Part II: The Domestic Story of the British Response to Climate Change Public Choice, Institutional Economics, Environmental Goods by *John O'Neill*
The International Financing of Environmental Protection: Lessons from Efforts to Protect the River Rhine against Chloride Pollution by *Thomas Bernauer*
Explaining National Variations of Air Pollution Levels: Political Institutions and Their Impact on Environmental Policy-Making by *Markus M L Crepaz*

ISSN 0964-0416 Volume 5 1996
Quarterly: Spring, Summer, Autumn, Winter
Individuals £42/$55 Institutions £115/$155

UK/OVERSEAS ORDERS to: Frank Cass, 890–900 Eastern Avenue,
Ilford, Essex IG2 7HH, UK. Tel: 0181 599 8866.
Fax: 0181 599 0984 E-mail: sales@frankcass.com
US ORDERS to: Frank Cass, c/o ISBS,
5804 N E Hassalo Street, Portland, OR 97213-3644, USA.
Tel: (503) 287-3093, (800) 944-6190 Fax: (503) 280-8832 E-mail:orders@isbs.com

A World Safe from Natural Disasters:
The Journey of Latin America and the Caribbean

The countries of Latin America and the Caribbean share histories of devastating natural disasters and, to a great extent, a present reality of social and economic insecurity. But in the face of sometimes daunting odds, these nations have developed the human resources and institutions necessary to cope with the effects of natural disasters and have strengthened their resolve to reduce the impact of future events on the well-being of their populations. The Pan American Health Organization and the IDNDR Secretariat, Regional Office for Latin America and the Caribbean, published *A World Safe from Natural Disasters: The Journey of Latin America and the Caribbean* as a contribution to the United Nations World Conference on Natural Disaster Reduction. The book is now avail-

able to the general public in English and Spanish at a cost of US$20.00.

With contributions from countless national officials and disaster management experts, and illustrated with color images from the PAHO/WHO archives, the book traces the transition over the past two decades from an era of improvised response and poorly coordinated international assistance to the more aggressive stance on disaster preparedness, mitigation, and prevention being adopted in many countries today.

To order this publication, please contact:

**Emergency Preparedness Program
PAHO/WHO
525 23rd. St., N.W.
Washington, DC 20037
Tel: (202) 861-6096
Fax: (202) 775-4578
Internet: disaster@paho.org**

Hacia un mundo más seguro frente a los desastres naturales:
La trayectoria de América Latina y el Caribe

Los países de América Latina y del Caribe comparten una historia de desastres naturales devastadores y hasta cierto punto una realidad actual caracterizada por la inseguridad social y económica. Sin embargo, estas naciones han hecho frente a unas disparidades a veces atemorizantes y han desarrollado los recursos humanos y las instituciones necesarias para sobrellevar los efectos de los desastres naturales al mismo tiempo que se han resuelto a reducir su impacto en el futuro para el bienestar de sus poblaciones.

La Organización Panamericana de la Salud y la Oficina Regional del DIRDN para América Latina y el Caribe, publicaron el libro *Hacia un mundo más seguro frente a los desastres naturales: La trayectoria de América Latina y el Caribe* como una contribución a la Conferencia Mundial para

la Reducción de Desastres Naturales que se celebró en 1994 en Yokohama, Japón. El libro ahora está disponible al público en inglés y español con un costo de US$20.00

Con contribuciones de innumerables oficiales nacionales y expertos en el manejo de desastres, e ilustrado con fotografías de la OPS, el libro traza la transición durante las últimas dos décadas desde una era de respuesta improvisada y asistencia internacional mal coordinada hasta una postura más agresiva sobre los preparativos para casos de desastre, su mitigación y la prevención que se está adoptando en muchos países hoy en día.

Para ordenar esta publicación por favor escriba a:

**Emergency Preparedness Program
PAHO/WHO
525 23rd. St., N.W.
Washington, DC 20037
Tel: (202) 861-6096
Fax: (202) 775-4578
Internet: disaster@paho.org**

CIIR Action for change

Publications

• Shadows behind the screen: Economic restructuring and Asian women

Combines personal life stories from women working in South China, Hong Kong, the Philippines, Thailand and Vietnam with country reports and statstics from Asia describing the impact of economic change on women's lives.
CIIR and ARENA (Hong Kong) 237 pages £7.99

Briefings & Reports

• Responding to the illegal drugs trade: in search of just and effective solutions

Calls for a reassessment of international drugs control policies. It assesses the most appropriate ways to tackle drug control, either by the law enforcement approach, which represses peasants in the Andes who grow coca, or the economic approach which would give them alternative livelihoods.
CIIR Drugs Trade & Development Series, ISBN 1852871563, 57 pages £2.00

• The European Union and the ASEAN

Shows that the EU appears to be making a shift away from development goals towards an economic agenda of free trade and open investment aims. This paper assesses what drives European policy and how policy decisions are made.
ISBN 1852871474, 40 pages £3.50

• Other CIIR briefings and reports:

Central America in the new world context, South Africa: Lessons from Zimbabwe, Sustainable or bankrupt: The Common Agricultural Policy, The United Nations & crisis management & Women & aid in the Philippines ...

Comment Series

• Haiti: Building democracy

Analyses the legacy of the 1991 coup and the return of Aristide. Looks at new challenges for the transition to democracy including: the changing role of the police force, human rights monitoring, economic emergency programmes and the role of civil society.
ISBN 1852871407 32 pages £2.50

• Mexico: Free market failure

Assesses the aftermath of Mexico's 1995 financial crisis. Examines the growing inequalities enshrined in NAFTA and the ruling Institutional Revolutionary Party's policies. Charts recent challenges to the political system by the Zapatista movement and civic action groups.
ISBN 1852871415 32 pages £2.50

• Other titles in the Comment series:
Coca, cocaine & the war on drugs, Babymilk, Biodiversity, the Philippines ...

• Forthcoming:
East Timor, South Africa & EU Coherence.

'At a time when others are withdrawing from the field because the South seems too far away or the problems at home too great, CIIR can be relied on to supply timely, accurate, succinct and usable information.'
Susan George, author & associate director of the Transnational Institute.

International Journal of
Mass Emergencies and Disasters

▲ *Journal* focuses on **Social** and **Behavioral** aspects of natural and technological threats
▲ Includes theory, research, planning, and policy
▲ Published triannually
▲ Official *Journal* of the Research Committee on Disasters, International Sociological Association, Thomas E. Drabek, President

Association Membership:
Kathleen Tierney
Disaster Research Center
University of Delaware
Newark DE 19716

Manuscripts:
Ronald W. Perry
School of Public Affairs
Box 870603
Arizona State University
Tempe AZ 85287-0603

Third World Quarterly

EDITOR
Shahid Qadir, *Royal Holloway, University of London, UK*

Supported by an International Editorial Board

Third World Quarterly is a leading journal of scholarship and policy in the field of international studies. For almost two decades it has set the agenda on Third World Affairs. As the most influential academic journal covering the emerging world, ***Third World Quarterly*** is at the forefront of analysis and commentary on fundamental issues of global concern.

Third World Quarterly provides expert and interdisciplinary insight into crucial issues before they impinge upon media attention, as well as coverage of the latest publications in its comprehensive book reviews section.

Third World Quarterly's original articles provide diverse perspectives on the developmental process and a candid discussion on democratic transitions, as well as identifying significant political, economic and social issues. Readable, free from esoteric jargon, informed without being abstruse, always authoritative and often provocative.

1996 - Volume 17. 5 issues. ISSN 0143-6597

FOR YOUR FREE SAMPLE COPY CONTACT:
Carfax Publishing Company
PO Box 25 • Abingdon • Oxfordshire OX14 3UE • UK
Tel: +44 (0) 1235 521154 • Fax: +44 (0) 1235 401550 • e-mail: tracyr@carfax.co.uk

warreport

WarReport provides in-depth coverage of key questions for the future of the Balkans, with its unique network of leading regional writers and analysts from the region. WarReport special features - on the UN, Bosnia, the Albanian National Question, Macedonian-Bulgarian relations, and others - provide in-depth studies and open debate on urgent issues of conflict and resolution.

"A voice of sanity amid the nationalist cacophony" - *The Guardian*
"WarReport is always of great value to us. The articles are always informative, well-argued, original and crisply edited" - A source, UNPROFOR

Subscription Rates: Individual rate: 6 issue subscription: £20/$30; 12 issue subscription: £30/$45; Institutional rate: 12 issue subscription: £50/$75; Supporter rate: 12 issue subscription: £100/$150. Concessionary rates available on request. Back issues are still available at £3.00, including:

Issue 39   Dayton: How the Bosnians Were Broken
Issue 38   Bosnia: Partition and Beyond
Issue 37   United Nations: With No Peace to Keep

Institute for War and Peace Reporting, Lancaster House, 33 Islington High St, London N1 9LH; Tel: 0171 713 7130; Fax: 0171 713 7140; email: warreport@gn.apc.org.

THE health EXCHANGE

The Health Exchange magazine brings you unique, first-hand coverage of the challenges facing health practitioners in developing countries. This lively and critical magazine reports on practical approaches to health development and explores the issues affecting people's health in low-income countries worldwide.

If you're looking for a job in a developing country *The Health Exchange* and Job Supplement are the first place to look. Every month international agencies publicise dozens of posts for health workers across the world. Advertisers include Oxfam, Save the Children Fund, the Overseas Development Administration and many others.

ihe
INTERNATIONAL
HEALTH EXCHANGE

To subscribe or for more
details, please mail, fax or call:
International Health Exchange
8-10 Dryden Street
London WC2E 9NA
Tel: +44 (0)171 836 5833
Fax: +44 (0)171 379 1239

Frank Cass

Small Wars and Insurgencies

Editors **Ian Beckett,** *University of Luton, UK* and
Thomas-Durell Young, *US Army War College,*
Pennsylvania, USA

Small Wars and Insurgencies is directed at providing a forum for
the discussion of the historical, political, social, economic and
psychological aspects of insurgency, counter-insurgency, limited
war, peace-keeping operations and the use of force as an
instrument of policy. Its aim is to provide an outlet for historians,
political scientists, policy makers and practitioners to discuss and
debate theoretical and practical issues related to the past, present
and future of this important area of both international and domestic
relations.

Recent Articles

**Two Faces of 1950s Terrorism: The Film Presentation of Mau Mau and the
Malayan Emergency** by *Susan Carruthers*
Peacekeeping in Intra-state Conflict by *Thomas Mockaitis*
Mexico in Crisis by *Donald E Schultz*
**An African Army Under Pressure: The Politicisation of the Malawi Army
and Operation Bwezani, 1992–1993** by *Jonathan Newell*
Army Helicopters, Navy Ships and Operation Earnest Will
by *Peter Clemens*
**The British Army's Development of Modern Counter-Guerrilla Warfare,
1944–1952** by *Tim Jones*

ISSN 0959-2318 Volume 7 1996
Three issues per year: Spring, Summer, Winter
Individuals £33/$48 Institutions £95/$145

UK/OVERSEAS ORDERS to: Frank Cass, 890–900 Eastern Avenue,
Ilford, Essex IG2 7HH, UK. Tel: 0181 599 8866
Fax: 0181 599 0984 E-mail: sales@frankcass.com
US ORDERS to: Frank Cass, c/o ISBS,
5804 N E Hassalo Street, Portland, OR 97213-3644, USA.
Tel: (503) 287-3093, (800) 944-6190 Fax: (503) 280-8832. E-mail: orders@isbs.com

Frank Cass

Low Intensity Conflict & Law Enforcement

Editor
Graham Hall Turbiville, Jr, *Fort Leavenworth, Kansas, USA*
European Editor
Mark Galeotti, *Keele University, UK*
Latin American Editor
Enrique Obando Arbulú, *Peruvian Center for International Studies, Lima, Peru*
Book Reviews Editor
John T Fishel, *Fort Leavenworth, Kansas, USA*

Low Intensity Conflict & Law Enforcement addresses a range of military, security, and law enforcement issues associated with conflict short of general war. More specifically, journal articles and features assess insurgency and counter-insurgency concepts and operations; the transnational threats of narcotics trafficking and terrorism in their many dimensions; peacekeeping and peacemaking operations; security assistance, disaster relief, and civil affairs programmes; special operations; and other security issues falling under the rubric of 'low intensity conflict.'

Low Intensity Conflict & Law Enforcement is essential reading for military and law enforcement professionals, and for scholars specializing in security issues.

Recent Articles

'Political Peacefare': A New United Nations Role? by *Monte R. Bullard* and *Patrick L Hatcher*
Facing the Unpalatable: The US Military and Law Enforcement in Operations Other Than War by *William Rosenau*
Human Rights and Counterinsurgency: Peruvian Army Officer Perspectives by *Richard W Shepard*
The Restructuring of the Security Services in Post-Communist Russia by *Dennis Desmond*
Raúl Castro, Revolutionary Cuba, and Institutional Stability by *Carlos G. Capplonch*
Internal Security; Terrorism, Government, Business, and the European Trends by Myles Robertson

ISSN 0966-2847 Volume 5 1996
Three issues per year: Summer, Autumn, Winter
Individuals £40/$50 Institutions £95/$135

UK/OVERSEAS ORDERS to: Frank Cass. 890–900 Eastern Avenue.
Ilford. Essex IG2 7HH. UK. Tel: 0181 599 8866
Fax: 0181 599 0984 E-mail: sales@frankcass.com
US ORDERS to: Frank Cass. c/o ISBS.
5804 N E Hassalo Street. Portland. OR 97213-3644. USA.
Tel: (503) 287-3093. (800) 944-6190 Fax: (503) 280-8832. E-mail:orders@isbs.com

AIDWATCH

AIDWATCH performs a signposting role on aid and poverty focused development. Quick to read and providing easy access to information, AIDWATCH covers news, books & conferences. It summarises new research, research in progress and regularly lists the contents of ten major development journals.

Published every 2 months, AIDWATCH is sent to over 2500 individual subscribers in 70 countries. Many organisations also distribute copies to staff or partner organisations.

Annual subscriptions cover 6 issues & include postage.

Single subscription .....£20   Five copies ......£30   10 copies.......£40
15 copies ....£50 25 copies.....£65 50 copies...£85 100 copies....£120

I enclose a cheque (£20 or $30) payable to Development Initiatives.

Please charge my credit card number (below)

...../...../...../...../...../...../...../...../...../...../...../...../...../...../...../.....

Expiry date   ...../..../...../.....   Signed...

Name..Address...

....

Fax.................................... Phone..

**Development Initiatives undertakes research and evaluation on aid, development policy, & NGO government relations. Contact: Old Westbrook Farm, Evercreech, Som, BA4 6DS, UK.
Tel +44 (0) 1749 831141 Fax +44 (0) 1749 831467
email: devinit@gn.apc.org**

The Australian Journal of Emergency Management

This Journal is produced by the Australian Emergency Management Institute (AEMI), the Training, Education and Research arm of Emergency Management Australia (EMA)

The Journal endeavours to provide an information sharing forum for all those involved in emergency management. Although its primary focus is on Australian issues and concerns, it also seeks to disseminate information from the international emergency/disasters community to enhance the development of an Australian capability within this field. The Journal is distributed without charge to subscribers in Australia and is sent to people and organisations throughout the world. If you wish to subscribe to this publication, please write or fax your request to the editor.

It is published quarterly. If you wish to contribute an article to this journal, please write to the editor for details. The address is:

The Editor, The Australian Journal of Emergency Management,
Information & Research Centre, Australian Emergency Management Institute, Mt Macedon, Victoria, 3441, Australia. Tel +61 54 215100, fax +61 54 215273.

Relief and Rehabilitation Network

The ***Relief and Rehabilitation Network*** (RRN) was launched in 1994 by the Overseas Development Institute, in conjunction with EuronAid, to provide a mechanism for the exchange of experience and good practice between individuals and organisations working in emergencies around the world. It aims to build institutional memory and professionalism among relief practitioners, and so increase the effectiveness of humanitarian interventions.

Since its launch, the RRN has grown to include nearly 300 members - 70% field-based - in over 50 countries. The Network offers access to practical, specialist information to personnel from NGOs, bilateral and multilateral organisations, the media and academic communities, as a means of bridging the gap between headquarters and field staff and between research and practice. Members currently receive four mailings per year, in English or French, comprising: Newsletters, Network Papers and Good Practice Reviews. Contributions from members are actively encouraged, as is feedback on publications received.

For more information, please contact:
The Relief and Rehabilitation Network
Overseas Development Institute
Regent's College, Regent's Park,
London NW1 4NS
Tel: +44 (0) 171 487 7591/7601 Fax: +44 (0) 171 487 7590 email: rrn@odi.org.uk.
More information is also available through the ODI Home Page on http://www.oneworld.org/odi

 **DISASTERS**

The Journal of Disaster Studies and Management

Edited by Charlotte Benson and Joanna Macrae, **Overseas Development Institute**

Disasters is the only journal that brings together research on disasters, vulnerability and relief and emergency management. The scope of the journal extends from disasters associated with natural hazards, such as earthquakes and drought, through to complex, conflict-related emergencies. It also contains conference reports and book reviews and welcomes correspondence and discussion. ***Disasters*** is published in March, June, September and December.

Subscription rates (Volume 20, 1996) are as follows:
Institutional Rates, £107.00 (UK-Europe), $172.00 (N.America), £107.00 (Rest of the World)
Personal Rates, £55.00 (UK-Europe), $103.00 (N.America), £64.00 (Rest of the World)
Reduced Rates, £45.00 (UK-Europe), $77.00 (N.America), £48.00 (Rest of the World)
(2 or more subscriptions at same address)

To receive an order form, please contact: Journals Marketing, Blackwell Publishers,
108 Cowley Road, Oxford OX4 1JF, UK. Tel: +44 (0) 1865 791100. Fax: +44 (0) 1865 791347.
Or to: Marketing, DISA, Blackwell Publishers, 238 Main Street, Cambridge, MA 02142, USA.
Tel: + 1 (617) 547 7100. Fax: +1 (617) 547 0789.

To apply for a free sample copy by email: jnlsamples@blackwellpublishers.co.uk

 **BLACKWELL** *Publishers*

International Centre for Humanitarian Reporting

Early, accurate and sustained press coverage of humanitarian crises and conflicts is vital. So is access to critical background reporting, humanitarian monitoring and contacts. For international relief professionals, policy makers, even the military.

The International Centre for Humanitarian Reporting (ICHR) and *CROSSLINES Global Report* seek to provide such reporting and background information.

International Centre for Humanitarian Reporting Headquartered in Geneva with a support office in Cambridge, MA. (USA), the ICHR stresses practical approaches to help decision makers, aid agencies, the military, and private sector better understand the roles of the media. Working closely with journalists but also relief professionals, private sector representatives and concerned individuals and institutions worldwide, the ICHR is in a unique position to facilitate more consistent, quality reporting of humanitarian issues through its Global Contact Network, conferences and workshops, journalist support fund, and radio outreach projects.

CROSSLINES Global Report
A leading newsjournal publishing 10 issues a year on international humanitarian relief, development and global trends, *CROSSLINES Global Report* seeks to provide more extensive background reporting by experienced journalists and also to serve as an independent forum for aid professionals and analysts.

Forthcoming issues in 1996 will focus on:

• Famine and food security
• Radio as public service to local populations in crises areas
• Child violence, prostitution and slavery
• Afghanistan
• The private sector in humanitarian relief
• Civilians in conflicts

CROSSLINES Special Reports

• *Somalia, Rwanda and Beyond - The Role of the International Media in Wars and Humanitarian Crises* (Currently available at $15/20 CHF plus P&P).

• *Weapons of War, Tools of Peace - The Role of the Media, Military and Private Sector in Wars and Humanitarian Crises* (To be published in the summer of 1996).

Support the International Centre for Humanitarian Reporting. Membership includes a subscription to *CROSSLINES Global Report* and 20% discount on other *CROSSLINES* publications.

• Individual Membership: Sfr.80/$65.
• Voluntary relief worker/Student: Sfr.50/$40.
• NGO and Non-Profit Organization Member: Sfr100/$85.
• Sponsorship Member: Sfr225/$150.
• Corporate/Institutional Member: Sfr300/$250.

Or simply subscribe to *CROSSLINES Global Report* ($45 for 10 issues in US and Europe). Group rates available.

Contacts:
International Centre for Humanitarian Reporting (ICHR).
5 chemin des Iris, CH-1216 Geneva, Switzerland.
Tel: 41(22)-920-1676. Fax: 41(22)-920-1679.
Email: info.ichr@itu.ch

CROSSLINES Global Report

US: Eve Porter. Tel:1(617)-491-4771.
Fax:1(617)-491-4689.
Email: Crosslines@aol.com.

Switzerland: Sue Pfiffner.
Tel/Fax: 41(22)-756-1984.

UK: Milly Taylor, Media Natura.
Tel: 44(171)-240-4936.
Fax: 44(171)-240-2291.

CROSSLINES GLOBAL REPORT

A Newsjournal on Humanitarian Action, Development and World Trends

Frank Cass

Terrorism and Political Violence

Editors **David C Rapoport,** *University of California, Los Angeles, USA* and **Paul Wilkinson,** *University of St Andrews, UK*

Terrorism and Political Violence reflects the full range of current scholarly work from many disciplines and theoretical perspectives. It aims to give academic rigour to a field which hitherto has lacked it, and encourages comparative studies. In addition to focusing on the political meaning of terrorist activity, the journal publishes studies of various related forms of violence by rebels and by states, on the links between political violence and organized crime, protest, rebellion, revolution, and human rights. A truly interdisciplinary journal, it is essential reading for all academics, decision makers and security specialists concerned with understanding political violence.

Recent Articles

Conflict Regulation or Conflict Resolution: Third Party Intervention in the Northern Ireland Conflict by *Sean Byrne*
Liberalism and the Limits of Pluralism by *Raphael Cohen-Almagor*
The Intifada: Palestinian Adaptation to Israeli Counterinsurgency Tactics by *Ruth Margolies Beitler*
Greece: Twenty Years of Political Terrorism by *George Kassimeris*
Colombia's Palace of Justice Tragedy Revisited by *Rex A Hudson*
Special Issues
Terror from the Extreme Right edited by *Tore Bjørgo*
Millennialism and Violence edited by *Michael Barkun*

ISSN 0954-6553 Volume 8 1996
Quarterly: Spring, Summer, Autumn, Winter
Individuals £42/$48 Institutions £135/$195

UK/OVERSEAS ORDERS to: Frank Cass. 890–900 Eastern Avenue. Ilford. Essex IG2 7HH. UK. Tel: 0181 599 8866
Fax: 0181 599 0984 E-mail: sales@frankcass.com
US ORDERS to: Frank Cass. c/o ISBS.
5804 N E Hassalo Street. Portland. OR 97213-3644. USA.
Tel: (503) 287-3093. (800) 944-6190 Fax: (503) 280-8832. E-mail:orders@isbs.com

Frank Cass

The Journal of Slavic Military Studies

(formerly The Journal of Soviet Military Studies)

Editors **David M Glantz,** *Carlisle, Pennsylvania, USA* and
Christopher Donnelly, *NATO HQ, Brussels, Belgium*

The Journal of Slavic Military Studies investigates all aspects of military affairs in the Slavic nations of central and eastern Europe in historical and geopolitical context and offers a vehicle for central and eastern European security and military analysts to air their views. Its unique international editorial board and diverse content including translations of newly released Soviet and Russian documents as well as specialist book reviews make the journal a must for academics, military figures and civilians alike who are interested in this region's security and military affairs.

Recent Articles

Prelude to Kursk–Soviet Strategic Operations, February-March 1943
by *David M Glantz*
The Armed Forces of the Baltic States: Current States and Problems of Development by *I Skrastins*
Russia's Defense Conversion: Theory Without Praxis by *Eugene Kogan*
The Russian Armed Forces Confront Chechnya, Military–Political Aspects and Military Activities, 11–31 December 1994 by *Timothy L Thomas*
Using Svechin to Analyze the Soviet Incursion into Afghanistan, 1979–1989 by *Scott E McIntosh*
Newly Published Soviet Works on the Red Army, 1918–1991
by *David M Glantz*

ISSN 1351-8046 Volume 9 1996
Quarterly: March, June, September, December
Individuals £42/$48 Institutions £130/$195

UK/OVERSEAS ORDERS to: Frank Cass. 890–900 Eastern Avenue.
Ilford. Essex IG2 7HH. UK. Tel: 0181 599 8866
Fax: 0181 599 0984 E-mail: sales@frankcass.com
US ORDERS to: Frank Cass. c/o ISBS.
5804 N E Hassalo Street. Portland. OR 97213-3644. USA.
Tel: (503) 287-3093. (800) 944-6190 Fax: (503) 280-8832. E-mail:orders@isbs.com

NEW INTERNATIONALIST MAGAZINE

Don't you get it?

Girls and girlhood
Time we were noticed

who's being exploited?

Green justice

The South speaks out

Filthy rich!
The new robber barons

who's making the money?

which stories don't make the news?

Get a grasp of the events and ideologies that shape an ever changing world. The NI magazine turns the inside out and explains what's really going on. It's the best guide to the major issues, from The Arms Trade to AIDS or Human Rights to Hunger. Each month the NI tackles one subject and gives you the facts and arguments so you have a clear understanding, and it's much quicker to read than a book. To influence what's happening you need to know what's going on.